NETWORK SCATTERING PARAMETERS

R Mavaddat
University of Western Australia

ADVANCED SERIES IN CIRCUITS AND SYSTEMS

Editor-in-Charge: **Wai-Kai Chen** (Univ. Illinois, Chicago, USA)
Associate Editor: **Dieter A. Mlynski** (Univ. Karlsruhe, Germany)

Published

Vol. 1: Interval Methods for Circuit Analysis
　　　　by *L. V. Kolev*

Advanced Series in Circuits and Systems – Vol. 2

NETWORK SCATTERING PARAMETERS

R Mavaddat

University of Western Australia

World Scientific

Singapore • New Jersey • London • Hong Kong

Published by

World Scientific Publishing Co. Pte. Ltd.
P O Box 128, Farrer Road, Singapore 912805
USA office: Suite 1B, 1060 Main Street, River Edge, NJ 07661
UK office: 57 Shelton Street, Covent Garden, London WC2H 9HE

ISBN 981-02-2305-6

Printed in Singapore.

Contents

5. Power Considerations of a 2-port Network

6. Generalized Scattering Parameters of a 2-port Network

7. The Scattering Parameters of an *N*-port Network

Preface

Network scattering parameters are powerful tools for the analysis and design of high frequency and microwave networks. Numerous books covering the theory and application of high frequency and microwave devices and circuits have been written and are available. These books include scattering parameters as appropriate tools for the analysis and design of the networks in these frequency ranges. However, broader in scope and covering a more comprehensive range of topics, careful consideration has not been given to many aspects of scattering parameters analysis. This book is an attempt to provide a clear and comprehensive review of network scattering parameters with detailed discussions of their application in the analysis of stability, input and output reflection coefficients, power gains and other circuit parameters. Numerous illustrative examples are given. There is presently no other book dealing exclusively with the application of scattering parameters in network analysis. It is hoped that the book will prove to be a useful companion to practicing engineers and valuable aid for students and teachers in the fields of high frequency, microwaves and optics.

I would like to express my gratitude to Mr. Brett Meyers for careful reading of the manuscript and checking of calculations, and for his many suggestions. I would also like to thank Ms. Linda Barbour for typing the manuscript. The use of 'Mathematica' for many calculations and graphical presentations is also acknowledged.

A short note on units

The units of ampere, volt, watt, ohm and siemens are used, respectively, for the current, voltage, power, resistance and conductance throughout the book. As these parameters are frequently repeated in the calculations, the units are not specified in every instance but implied.

Chapter 1

Introduction and the Scattering Parameters of a 1-port Network

The present Chapter serves as an introduction to the network scattering parameters. To illustrate the concept and application of the scattering parameters, the simplest case of a 1-port network is discussed and the network stability and the transfer of power from generator to load are considered. The scattering parameters of a network are defined in terms of a set of reference impedances. If these impedances are not purely resistive, the scattering parameters are known as the generalized scattering parameters. The generalized scattering parameters are briefly introduced and the generalized scattering parameters of 1-port network examined.

1.1 Introduction

The scattering parameters analysis developed in this book is a steady state analysis applicable to any network and at any frequency. At lower frequencies, however, other circuit matrix representations, such as impedance or admittance matrices, are equally applicable and the use of scattering parameters is not essential. The application of the scattering parameters is mainly in the microwave and optical frequency regions of the electromagnetic spectrum. At these higher frequencies, although it is possible to define certain equivalent voltages and currents and represent networks by their impedance and admittance parameters, this is not appropriate for the following reasons:

(1) It is not always possible to define appropriate voltages or currents for a network in a unique manner.

(2) Direct measurements of voltages and currents and hence impedances and admittances are not always practical.

(3) Measurements of the impedance and admittance parameters, if possible, are to be performed under short circuit or open circuit conditions, not easily achieved in practice at higher frequencies. In many cases such conditions lead to circuit oscillation.

In this section we briefly define and discuss the network scattering parameters and explain how the use of these parameters eliminates the above mentioned problems related to other circuit representations.

In general, a circuit consists of any number of sources, guiding structures, passive or active multi-port networks and terminations. A typical circuit is shown schematically in Fig. (1.1.1) for a single *N*-port network.

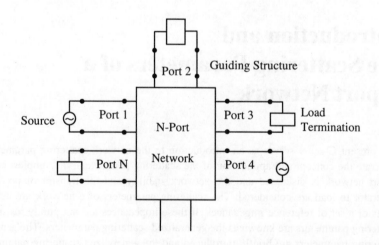

Fig. (1.1.1)

Sources of electromagnetic generations can take numerous forms with their frequencies ranging through the electromagnetic frequency spectrum, from low radio frequencies to optical frequencies and beyond. The generation of an electromagnetic field is usually at a center frequency with very small frequency variations (effect of modulation or noise), which are neglected in our discussion. Any generator has its own equivalent internal impedance.

Guiding structures could be sections of two-wire lines at low frequencies, parallel plates or coaxial lines at lower microwave frequencies, striplines or microstrip lines or rectangular or circular hollow conductor waveguides at microwave or millimeter wavelengths and finally planar dielectric and fibres at optical frequencies. Each of the above could have different geometrical dimensions and forms. The above list is not complete and frequencies of application may overlap.

The load termination is considered generally to be a device that totally or partially absorbs or utilizes the electric power generated by the source. The power not absorbed by the load is either dissipated within the network or reflected back to the source. Occasionally these terminations may reflect back all the received power and be termed reactive loads.

The N-port networks themselves may have numerous forms with very many applications. Some examples of these are the networks used for attenuation or amplification or as filters, impedance transformers, directional couplers, etc. In general the networks connect sources via the guiding structures to the load terminations.

In this book we are not concerned with the specific types of generators, guiding structures, networks or terminations. Rather, we are concerned with developing a general tool for analysis and understanding of the circuits thus constructed. Specifically we use the network scattering parameters for this analysis. Numerous other books can be consulted for specific networks and their applications. We also assume that the network is linear or the superimposed wave amplitudes (to be defined) can simply be added.

The network scattering parameters, can be developed purely on lumped or distributed circuit concepts, assigning emf and internal impedances to the generators, characteristic impedances to the guiding structures and impedances to the load terminations. However, it is not always possible to define or measure the voltages and currents, and hence impedances or admittances, in a unique way. In addition a guiding structure may support electromagnetic fields, in a number of waveguide modes. In general each mode carries a wave of different amplitude with its own spatial field configuration and phase and group velocities. Modes propagating in a single waveguide may also differ in the polarization of their transverse electric field. For each of these modes we should define a new characteristic impedance and phase constant.

The scattering parameters analysis has the advantage that in this analysis the actual values of the impedances are not relevant, but their effect on so called source or load reflection coefficients. The reflection coefficients can easily be measured.

For illustration consider a transmission line of characteristic impedance R_0, terminated by a load of impedance Z_L. A wave traveling along the transmission line is partially reflected at the load with a reflection coefficient given by

$$\Gamma_L = \frac{Z_L - R_0}{Z_L + R_0}. \tag{1.1.1}$$

Parameters similar to the above load reflection coefficient Γ_L that can be measured are relevant to the scattering parameters analysis, and not the actual values of the load impedance Z_L or the characteristic impedance R_0. Similar to the load reflection coefficient, we can define a source reflection coefficient Γ_S as

$$\Gamma_S = \frac{Z_S - R_0}{Z_S + R_0} \tag{1.1.2}$$

where Z_S is the impedance of the source.

The reflection coefficients are defined in this book as the ratio of the wave amplitudes of two coherent propagating waves, traveling in opposite directions. In the scattering parameters analysis, we can generalize this concept and assume that the relative wave amplitudes (magnitude and phase) of two waves propagating in one or different waveguides can be measured. The reflection coefficient is a special case when two waves of the same mode travel in opposite directions in a single waveguide.

Another parameter of importance that can also be measured is power. This can be the generated power by the source, power flow through a guiding structure or the absorbed power in the load. In the scattering parameters analysis, therefore, we are concerned basically with the relative wave amplitudes and relevant power terms. The use of voltages and currents, impedances and admittances are for analogy to lower frequency circuits and for convenience and are not essential.

We can now proceed with the definition of network scattering parameters. Consider a network arrangement shown in Fig. (1.1.1). For the sake of the scattering parameter definition and measurement, each propagating mode is assumed to be connected to the network by a single port. Consequently, the number of ports may exceed the number of actual physical connections to a network.

For defining the scattering parameters, we assume that sources and terminations are available that are *matched* to a particular guiding structure. Hence if the source is connected by any section of the guiding structure to the load termination, all power generated by the source is absorbed by the load termination. Both the source and the load are '*matched terminated*'.

For the network of Fig. (1.1.1), the scattering parameter S_{jj} is defined as the ratio of the reflected wave to that of the incident wave for port j, when all other ports are matched terminated. The scattering parameter S_{kj} is the ratio of the wave amplitude of the wave transmitted through port k, to that of the wave amplitude incident on port j, when all ports apart from the jth port are matched terminated. N^2 parameters so defined constitute the N by N network scattering matrix denoted here by **S**. Hence

$$\mathbf{S} = \begin{bmatrix} S_{11} & S_{12} & \cdot & S_{1j} & S_{1k} & \cdot & S_{1N} \\ S_{21} & S_{22} & \cdot & S_{2j} & S_{2k} & \cdot & S_{2N} \\ \cdot & & \cdot & & & \cdot & \cdot \\ S_{j1} & S_{j2} & \cdot & S_{jj} & S_{jk} & \cdot & S_{jN} \\ S_{k1} & S_{k2} & \cdot & S_{kj} & S_{kk} & \cdot & S_{kN} \\ \cdot & & \cdot & \cdot & \cdot & \cdot & \cdot \\ S_{N1} & S_{N2} & \cdot & S_{Nj} & S_{Nk} & \cdot & S_{NN} \end{bmatrix} \tag{1.1.3}$$

and

$$\mathbf{b} = \mathbf{S}\,\mathbf{a} \tag{1.1.4}$$

where **a** and **b** denote the wave amplitudes of the waves traveling towards the network and away from the network (reflected) respectively.

It is now clear that in defining the scattering parameters: (1) no reference is made to voltages, currents, impedances or admittances, (2) there is no need for the measurement of these parameters, and (3) open or short circuited terminations are not used in our definition. In spite of this discussion, we shall make extensive use of voltages and currents and admittance and impedance parameters. This is because these parameters should be more familiar to the reader and it is also a convenient way of introducing the idea of the scattering parameters.

The generalized scattering parameters

To define the scattering parameters, we have used sources and load terminations that were matched to the relevant guiding structures with resistive characteristic impedances. These resistive impedances are the reference impedances for the definition of the scattering parameters. It is possible to extend the definition of the scattering parameters to situations where other reference impedances are used. Specifically the reference impedances can be taken as the actual source and load impedances of the circuit, which may differ from the impedances of the guiding structures. These are sometimes convenient for writing abbreviated expressions or simplifying certain calculations.

In Chapters 2 to 5 our discussion is confined to the scattering parameters with resistive references. The generalized scattering parameters for a 2-port network and N-port networks are discussed in Chapters 6 and 7, respectively. In the way of introduction, the last section of this Chapter is devoted to the generalized scattering parameters of a 1-port network. This section, however, can be read later and in conjunction with the last two Chapters.

Although the scattering parameters are not uniquely defined, the measurement of these parameters is only convenient when the source and load terminations are matched to the guiding structures. For this reason, the scattering parameters in terms of the resistive reference impedances, equal to the matched impedances of the guiding structures, can be termed the measured scattering parameters.

The general plan of the book

The present Chapter serves as an introduction. We continue by illustrating the concept of scattering parameters by considering the scattering parameters of the simplest network, namely, the 1-port network. The 2-port network scattering parameters are introduced in Chapter 2. An important property of a network is its impedance transforming properties. This concept and the concept of impedance matching, using the 2-port network scattering parameters, are discussed in Chapter 3. The stability of 2-port networks is considered in Chapter 4, in terms of network scattering parameters. The 2-port network power and power gains together with unilateral and bilateral circles of constant power gain are discussed in Chapter 5. In Chapter 6 we introduce the 2-port network generalized scattering parameters and briefly discuss the network properties in terms of these generalized parameters. Chapter 7 extends the analysis to N-port networks.

1.2 Scattering Parameters of a 1-port Network

As explained in the last section, the scattering parameters of an N-port network are defined in terms of a set of N reference impedances. Thus a 1-port network requires a single reference impedance for its definition. Denoting this reference impedance by $\hat{Z}_1$, we presently assume that $\hat{Z}_1$ is resistive and is equal to R_{01}, the impedance of the waveguide (or transmission line) to be connected to the only port of the network.

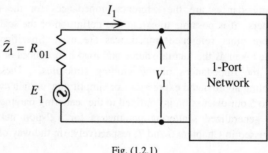

Fig. (1.2.1)

Current and voltage scattering parameters

To define the scattering parameter of a 1-port network we consider a generator of emf E_1 and internal impedance equal to the resistive reference impedance R_{01}, connected to the given 1-port network as shown in Fig. (1.2.1). We denote the current through and voltage across the network by I_1 and V_1 respectively.

To proceed, we first assume that the generator is disconnected from the network and connected to a load of impedance R_{01} (the reference impedance), as shown in Fig. (1.2.2). The current through and voltage across the impedance R_{01} are defined as the incident current and incident voltage and are denoted as I^i_1 and V^i_1 respectively.

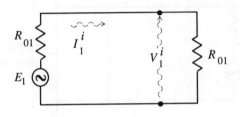

Fig. (1.2.2)

The incident current I^i_1 and voltage V^i_1 in terms of the total current I_1 and voltage V_1 can be written as

$$I^i_1 = \frac{E_1}{2R_{01}} = \frac{V_1 + R_{01}I_1}{2R_{01}}, \qquad (1.2.1)$$

$$V^i_1 = \frac{E_1}{2} = \frac{V_1 + R_{01}I_1}{2}. \qquad (1.2.2)$$

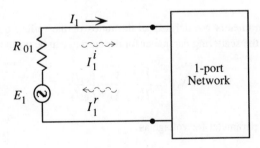

Fig. (1.2.3)

Next we replace the termination by the actual 1-port network as in Fig. (1.2.3). We define the reflected current $I^r{}_1$ by the expression

$$I^r{}_1 = I^i{}_1 - I_1.$$

Similarly by referring to Fig. (1.2.4) we define the reflected voltage as

$$V^r{}_1 = V_1 - V^i{}_1.$$

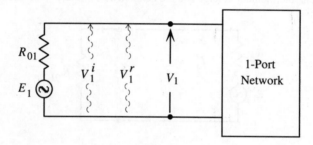

Fig. (1.2.4)

In terms of the total currents and voltages

$$I^r{}_1 = \frac{V_1 + R_{01}I_1}{2R_{01}} - I_1 = \frac{V_1 - R_{01}I_1}{2R_{01}}, \qquad (1.2.3)$$

$$V^r{}_1 = V_1 - \frac{V_1 + R_{01}I_1}{2} = \frac{V_1 - R_{01}I_1}{2}. \qquad (1.2.4)$$

The scattering parameters are defined as the ratios of the reflected to that of incident quantities. Hence the scattering parameter for the current can be written as

$$S_{11}{}^I = \frac{I^r{}_1}{I^i{}_1} = \frac{V_1 - R_{01}I_1}{V_1 + R_{01}I_1} \qquad (1.2.5)$$

and the scattering parameter for voltage as

$$S_{11}{}^V = \frac{V^r{}_1}{V^i{}_1} = \frac{V_1 - R_{01}I_1}{V_1 + R_{01}I_1}. \qquad (1.2.6)$$

In (1.2.5) and (1.2.6) we have used the notation $S_{11}{}^I$ and $S_{11}{}^V$ with double subscripts to conform to the similar notations used for a multi-port network.

The wave amplitude scattering parameter

From (1.2.5) and (1.2.6), the scattering parameters for the voltage and current are clearly equal. This is only true for this particular choice of reference impedance. For a different reference impedance the two scattering parameters would be proportional, rather than equal. In either case it is not necessary to define separate current and voltage scattering parameters, as these are readily convertible. Instead we define a single scattering parameter as the ratio of the, so called, *reflected wave amplitude* to that of the *incident wave amplitude* as defined below.

We define the incident wave amplitude as

$$a_1 = \sqrt{R_{01}}\, I^i{}_1 = \frac{1}{\sqrt{R_{01}}} V^i{}_1 = \frac{V_1 + R_{01} I_1}{2\sqrt{R_{01}}} \tag{1.2.7}$$

and the reflected wave amplitude as

$$b_1 = \sqrt{R_{01}}\, I^r{}_1 = \frac{1}{\sqrt{R_{01}}} V^r{}_1 = \frac{V_1 - R_{01} I_1}{2\sqrt{R_{01}}} \tag{1.2.8}$$

and hence the wave amplitude scattering parameters as

$$S_{11} = \frac{b_1}{a_1} = \frac{I^r{}_1}{I^i{}_1} = \frac{V^r{}_1}{V^i{}_1} = \frac{V_1 - R_{01} I_1}{V_1 + R_{01} I_1}. \tag{1.2.9}$$

For this particular choice of reference impedance, therefore, the new scattering parameter is equal to the current and voltage scattering parameters defined before, or

$$S_{11} = S_{11}{}^I = S_{11}{}^V. \tag{1.2.10}$$

Example (1.2.1)

Find the current, voltage and the wave amplitude scattering parameters for a 1-port network of impedance $Z_{11} = 75 + j100$ (double subscript is used to conform to the notations for an N-port network). Take the reference impedance $\hat{Z}_1 = R_{01} = 50$.

Solution

From (1.2.5) and with $V_1 = Z_{11}I_1$, the scattering parameter for current can be written as

$$S_{11}{}^I = I^r{}_1 / I^i{}_1 = (Z_{11} - R_{01}) / (Z_{11} + R_{01})$$

$$= (75 + j100 - 50) / (75 + j100 + 50) = 0.512 + j0.390.$$

Similarly from (1.2.6), the scattering parameter for voltage is

$$S_{11}{}^V = V^r{}_1 / V^i{}_1 = (Z_{11} - R_{01}) / (Z_{11} + R_{01}) = 0.512 + j0.390.$$

Finally from (1.2.10), $S_{11} = S_{11}{}^I = S_{11}{}^V = 0.512 + j0.390.$

1.3 The Wave Amplitudes of a General 1-port Network - Signal Flow Graphs

In the previous section we defined the incident and reflected wave amplitudes and the wave amplitude scattering parameter for a 1-port network. To define the incident and reflected wave amplitudes, we considered a generator with a source impedance equal to that of the reference impedance. In general the source impedance is different from the reference impedance and the incident and reflected wave amplitudes are modified. We should emphasize, however, that the modified wave amplitudes are still dependent on the reference impedance R_{01} for their definitions.

To see how the incident wave amplitude is modified, consider a 1-port network connected to a generator of emf E_S and actual impedance Z_S as shown in Fig. (1.3.1).

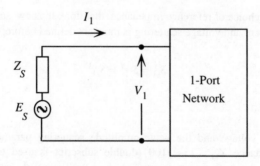

Fig. (1.3.1)

For this situation the incident wave amplitude a_1 can be considered as consisting of two parts, denoted by $a_1^{[1]}$ and $a_1^{[2]}$ with

$$a_1 = a_1^{[1]} + a_1^{[2]}.$$

The term $a_1^{[1]}$ is the incident wave amplitude when the network is replaced by a load equal to the reference impedance as in Fig. (1.3.2a).

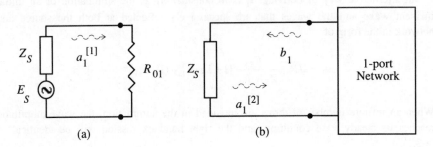

Fig. (1.3.2)

As in the previous section $a_1^{[1]}$ can be written as

$$a_1^{[1]} = \sqrt{R_{01}} \frac{E_S}{Z_S + R_{01}} = \frac{E_S}{2\sqrt{R_{01}}}(1 - \Gamma_S)$$

where

$$\Gamma_S = \frac{Z_S - R_{01}}{Z_S + R_{01}}$$

is the reflection coefficient of the source.

In addition to the above term, we have a contribution from the wave amplitude b_1 reflected at the source as shown in Fig. (1.3.2b). This additional contribution can be written as

$$a_1^{[2]} = \Gamma_S b_1.$$

Hence the modified expression for the incident wave a_1 can be written implicitly as

$$a_1 = \frac{E_S}{2\sqrt{R_{01}}}(1 - \Gamma_S) + \Gamma_S S_{11} a_1$$

where $S_{11} = b_1 / a_1$ is the network scattering parameter (independent of the source impedance) as defined before.

Solving for a_1 we have

$$a_1 = \frac{E_S}{2\sqrt{R_{01}}} \frac{1 - \Gamma_S}{1 - \Gamma_S S_{11}}. \tag{1.3.1}$$

We note that with $Z_S = R_{01}$, $\Gamma_S = 0$ and $a_1 = E_S / 2\sqrt{R_{01}}$ as in (1.2.7).

An alternative way of deriving a_1 is to consider a_1 as the summation of an initial incident wave and the waves that are successively reflected at both the source and network in the form of

$$a_1 = \sqrt{R_{01}} \frac{E_S}{Z_S + R_{01}} [1 + \Gamma_S S_{11} + (\Gamma_S S_{11})^2 + \cdots \cdots].$$

When an infinite number of terms are included in the summation, the wave amplitude reaches its steady state condition, and the right hand expression will be identical to (1.3.1).

Defining

$$b_S = \frac{E_S}{2\sqrt{R_{01}}} (1 - \Gamma_S) \tag{1.3.2}$$

the wave amplitudes can be written finally as

$$a_1 = \frac{b_S}{1 - \Gamma_S S_{11}}, \tag{1.3.3}$$

$$b_1 = S_{11} a_1. \tag{1.3.4}$$

The signal flow graph

The relations (1.3.3) and (1.3.4) can be represented by a flow graph as shown in Fig. (1.3.4). Each node in the diagram represents a wave amplitude (incident or reflected) and each directed branch the contribution to the node amplitude from an adjacent node.

With these conventions and the inspection of the flow diagram, we can write

$$a_1 = b_S \times (1) + b_1 \Gamma_S = b_S + b_1 \Gamma_S,$$

$$b_1 = a_1 S_{11}.$$

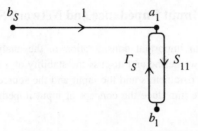

Fig. (1.3.4)

The second equation above is identical to (1.3.4), while eliminating b_1 between the two equations leads to expressions (1.3.3).

Example (1.3.1)

A source of emf $E_S = 5$ and impedance $Z_S = 25 - j50$, is connected to a network of impedance $Z_{11} = 75 + j100$. Find the amplitudes of the incident and reflected waves in a system of reference impedance $\hat{Z}_1 = R_{01} = 50$.

Solution

The source reflection coefficient can be found as

$$\Gamma_S = (Z_S - R_{01}) / (Z_S + R_{01}) = (-25 - j50) / (75 - j50) = 0.077 - j0.615.$$

With the reference impedance $\hat{Z}_1 = R_{01} = 50$, the scattering parameter

$$S_{11} = (Z_{11} - R_{01}) / (Z_{11} + R_{01}) = (25 + j100) / (125 + j100) = 0.512 + j0.390.$$

Hence from (1.3.2)

$$b_S = \frac{E_S}{2\sqrt{R_{01}}}(1 - \Gamma_S) = 0.326 + j218$$

and from (1.3.3) and (1.3.4), the incident and reflected wave amplitudes are given as

$$a_1 = \frac{b_S}{1 - S_{11}\Gamma_S} = 0.495 + j0.106,$$

$$b_1 = S_{11} a_1 = (0.512 + j0.390)(0.495 + j0.106) = 0.212 + j0.247.$$

1.4 1-Port Network Input Impedance and Network Stability

The circuit stability is an important consideration in the analysis and design of any network. In this section we define and discuss the stability of a 1-port network in terms of the network scattering parameters and the input and the source reflection coefficients. Prior to this, however, we introduce the concept of input impedance and define passive and active 1-port networks.

Input impedance and input reflection coefficient

The impedance of a 1-port network seen at its input can be termed the input impedance of the network and denoted by Z_{IN}. With the current across the port denoted by I_1 and voltage by V_1, we have

$$Z_{IN} = \frac{V_1}{I_1} = Z_{11}$$

(In general for the case of an N-port network $Z_{IN} \neq Z_{11}$).

We can define the input reflection coefficient by an expression similar to (1.1.1) as

$$\Gamma_{IN} = \frac{Z_{IN} - R_{01}}{Z_{IN} + R_{01}}. \tag{1.4.1}$$

From (1.2.9) it is clear that for a 1-port network

$$\Gamma_{IN} = S_{11}. \tag{1.4.2}$$

The input impedance can be written in terms of the input reflection coefficient as

$$Z_{IN} = \frac{1 + \Gamma_{IN}}{1 - \Gamma_{IN}} R_{01} = \frac{1 + S_{11}}{1 - S_{11}} R_{01}. \tag{1.4.3}$$

Passive and active 1-port networks

A 1-port network is said to be passive, if the resistive part of its input impedance is zero or positive. The network is active if the resistive part of its impedance is negative. Hence for a network to be passive we require

$$\text{Re}(Z_{IN}) \geq 0.$$

The above equation can be written as

$$Z_{IN} + Z_{IN}^* \geq 0$$

or from (1.4.3), in terms of the scattering parameters

$$\left[\frac{1+S_{11}}{1-S_{11}} + \frac{1+S_{11}^*}{1-S_{11}^*}\right]R_{01} = \frac{2(1-S_{11}S_{11}^*)}{|1-S_{11}|^2} \geq 0.$$

For a 1-port network to be passive, therefore, we require

$$|S_{11}| \leq 1. \tag{1.4.4}$$

When $|S_{11}| = 1$, $Z_{IN} + Z_{IN}^* = 0$ and the network, as seen at its input, is purely reactive.

Network stability

Consider a 1-port network of input impedance Z_{IN} connected to a generator of internal impedance Z_S as shown in Fig. (1.4.1).

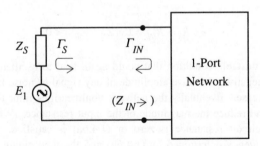

Fig. (1.4.1)

In general Z_S and Z_{IN} can both have resistive and reactive parts and can be written, respectively, as

$$Z_S = R_S + jX_S$$

and

$$Z_{IN} = R_{IN} + jX_{IN}.$$

Hence the expression

$$Z_S + Z_{IN} = 0 \qquad (1.4.5)$$

can be written as

$$R_S + R_{IN} = 0 \qquad (1.4.6a)$$

and

$$X_S + X_{IN} = 0. \qquad (1.4.6b)$$

It can now be seen that (1.4.5) expresses the condition for oscillation. The real part of (1.4.5), as given by (1.4.6a), ensures that any oscillations introduced will be sustained (there is no power dissipation), and its imaginary part as given by (1.4.6b) defines the oscillation frequency.

If a circuit is not oscillatory, it is said to be stable if

$$R_S + R_{IN} = \text{Re}(Z_S + Z_{IN}) > 0. \qquad (1.4.7)$$

On the other hand with

$$R_S + R_{IN} = \text{Re}(Z_S + Z_{IN}) < 0, \qquad (1.4.8)$$

the circuit has an unstable condition that could again lead to oscillation. With the net circuit resistance negative, the wave amplitude of any signal or noise introduced into the circuit, would increase. Eventually the network nonlinearities limit the increase of the wave amplitude and reduce the magnitude of the input resistance. When steady state is reached, the net circuit resistance is zero or (1.4.6a) is satisfied. The circuit then oscillates at a frequency determined by (1.4.6b) and the new circuit conditions. This frequency may differ from the frequency at the onset of the instability. In terms of the reflection coefficients the condition for a circuit to be oscillatory can be written as

$$\frac{\Gamma_S - 1}{\Gamma_S + 1} + \frac{S_{11} - 1}{S_{11} + 1} = 0$$

which simplifies to

$$1 - \Gamma_S S_{11} = 0. \qquad (1.4.9)$$

This is consistent with the expression (1.3.1) given for the wave amplitude as

$$a_1 = \frac{E_S}{2\sqrt{R_{01}}} \frac{1-\Gamma_S}{1-\Gamma_S S_{11}}.$$

When (1.4.9) is satisfied, the denominator of the above expression is zero, or the wave amplitude can be maintained with no generator input. This is the condition for a circuit to be oscillatory.

Unconditional stability or network stability for all source impedances

The stability condition (1.4.7) is for a particular value of the source resistance. In general, the network may be stable for some and potentially unstable for other values of the source resistance. We may, however, define a 1-port network as being *unconditionally stable* if it is stable for every value of the source resistance.

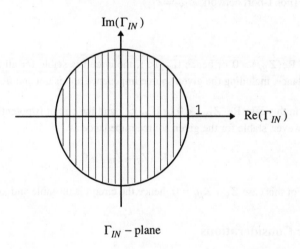

$$\Gamma_{IN} - \text{plane}$$

Fig. (1.4.2) - The shaded area is unconditionally stable.

For a network to be unconditionally stable we require the condition

$$\text{Re}(Z_{IN}) > 0 \tag{1.4.10}$$

to be satisfied.

In terms of the input reflection coefficient or the network scattering parameter, (1.4.10) can be written as

$$|\Gamma_{IN}| = |S_{11}| < 1. \tag{1.4.11}$$

We can present $|\Gamma_{IN}| = 1$ as a circle of unit radius in a complex Γ_{IN} plane as shown in Fig. (1.4.2). All load impedances with reflection coefficients Γ_{IN} inside this circle are passive and in conjunction with the source (passive source) create an unconditionally stable circuit. If however, Γ_{IN} is outside this circle, the circuit is potentially unstable. Presentations similar to this will prove very useful in the study of the N-port network stability conditions.

Example (1.4.1)

A generator of impedance $Z_S = 70 - j30$ is connected to a 1-port network of impedance Z_{11}, discuss the stability of the circuit if (i) $Z_{11} = 30 + j80$, (ii) $Z_{11} = -50 - j10$, (iii) $Z_{11} = -70 + j30$. (For 1-port network $Z_{IN} = Z_{11}$).

Solution

(i) As $\text{Re}(Z_{IN}) > 0$ or hence $|\Gamma_{IN}| < 1$, the circuit is stable for all values of source impedance, including the given source impedance and is unconditionally stable.

(ii)   In this case $\text{Re}(Z_{IN}) < 0$ or $|\Gamma_{IN}| > 1$ and the circuit is potentially unstable. It is, however, stable for the given source impedance as

$$\text{Re}(Z_S + Z_{IN}) > 0.$$

(iii) For this case $Z_S + Z_{IN} = 0$, hence the circuit is unstable and oscillates.

1.5 Power Considerations

In this section we define several power terms and give an expression for the power gain of a 1-port network. Although these definitions are trivial for a 1-port Network, they are given for comparison with the power terms to be defined later for a multi-port network.

Power available from the source

The power available from the source is given as

$$P_{AVS} = e_S{}^2 \tag{1.5.1}$$

where

$$e_S = \frac{E_S}{2\sqrt{R_S}}.$$

In (1.5.1) we have taken the phase of E_S as the reference and set it equal to zero.

Incident and reflected powers

When a 1-port circuit is not oscillating, we define the term $a_1 a_1^*$ as the *incident power* and similarly $b_1 b_1^*$ as the *reflected power*.

Input power

The difference between the incident and reflected powers is the input power P_{IN} to the network and given as

$$P_{IN} = a_1 a_1^* - b_1 b_1^*. \tag{1.5.2}$$

For a 1-port network, this power is dissipated within the network itself.

The expression (1.5.2) can be readily verified for the case of $Z_S = R_{01}$ by substituting for a_1 and b_1, as given respectively by (1.2.7) and (1.2.8). This gives

$$P_{IN} = a_1 a_1^* - b_1 b_1^* = \frac{1}{2}(I_1 V_1^* + I_1^* V_1)$$

which is clearly the power delivered to the network. It is not difficult to show that the above relation also holds for all other source impedances.

The input power P_{IN} can be written in terms of b_S, Γ_S and the scattering parameter S_{11} as

$$P_{IN} = a_1 a_1^* - b_1 b_1^* = a_1 a_1^* (1 - |S_{11}|^2) = |b_S|^2 \frac{(1 - |S_{11}|^2)}{|1 - \Gamma_S S_{11}|^2}. \tag{1.5.3}$$

Substituting for b_S in terms of e_S as

$$b_S = (1 - \Gamma_S)\sqrt{\frac{1 - \Gamma_S \Gamma_S^*}{(1 - \Gamma_S)(1 - \Gamma_S^*)}}\, e_S,$$

we can finally write an expression for P_{IN} in terms of e_S as

$$P_{IN} = e_S^2 \frac{(1 - |\Gamma_S|^2)(1 - |S_{11}|^2)}{|1 - \Gamma_S S_{11}|^2}. \tag{1.5.4}$$

When $Z_{11} = Z_{IN} = Z_S^*$, we have $S_{11} = \Gamma_{IN} = \Gamma_S^*$ and hence $P_{IN} = e_S^2 = P_{AVS}$, as required.

The above expressions for power do not explicitly include any impedance terms. The input power term can hence be expressed in terms of the network scattering parameter, the source reflection coefficient and the power available from the source. This explains the inclusion of $\sqrt{R_{01}}$ terms in the definition of the wave amplitudes in (1.2.7) and (1.2.8).

Power gain

The ratio of the power dissipated in the network and the available power can be termed as the power gain of the circuit. Hence the power gain G_T is given by

$$G_T = \frac{P_{IN}}{P_{AVS}} = \frac{(1 - |\Gamma_S|^2)(1 - |S_{11}|^2)}{|1 - \Gamma_S S_{11}|^2}. \tag{1.5.5}$$

When the network is passive, $|S_{11}| < 1$ and the power gain is always less than unity.

Example (1.5.1)

A source of emf $E_S = 5$ and impedance $Z_S = 25 - j50$, is connected to a network of impedance $Z_{IN} = 75 + j100$. Find the power available from the source and dissipated in the load.

Solution

The power available from the source is given by

$$P_{AVS} = e_S^2 = \frac{E_S^2}{4R_S} = 0.25.$$

Using (1.5.4) and with $S_{11} = 0.512 + j0.390$ and $\Gamma_S = 0.077 - j0.615$, we find

$$P_{IN} = 0.25\left[\frac{(1 - 0.385)(1 - 0.415)}{0.600}\right] = 0.15.$$

To verify the above derivation, we can calculate the power from the usual circuit consideration. This is given as

$$P_{IN} = \text{Re}(V_1 I_1^*) = \text{Re}\left[\frac{5(75 + j100)}{100 + j50} \times \frac{5}{100 - j50}\right] = 0.15.$$

1.6 The Generalized Scattering Parameters

In this section we consider the representation of a 1-port network in terms of the generalized scattering parameters and derive the conversion relations between the measured scattering parameters and the generalized scattering parameters. The generalized scattering parameters are not used in the discussion of 2-port networks in Chapters 2 to 5. Hence this section can be read in conjunction with the last two chapters of this book.

As explained in the introduction, the scattering parameters are defined in terms of a set of reference impedances. The scattering parameter of a 1-port network requires a single reference impedance for its definition. So far it has been assumed that this reference impedance $\hat{Z}_1$ is resistive and was denoted by $R_{01}(R_{01}$ was defined previously). In general, however, $\hat{Z}_1$ could be assumed to be a complex impedance, different from the resistive reference impedance R_{01}. To find the conversion between the two scattering parameters, with different reference impedances, it is convenient to define a reflection coefficient γ_1 in terms of the old and the new reference impedances in the form of

$$\gamma_1 = \frac{\hat{Z}_1 - R_{01}}{\hat{Z}_1 + R_{01}}. \tag{1.6.1}$$

This can be called the reference reflection coefficient. The reference impedance $\hat{Z}_1$ is given in terms of the reference reflection coefficient γ_1 as

$$\hat{Z}_1 = \frac{1 + \gamma_1}{1 - \gamma_1} R_{01}. \tag{1.6.2}$$

The reference impedance $\hat{Z}_1$ and reference coefficient γ_1 can be used interchangeably in the definition of the scattering parameters. With $\mathrm{Re}(\hat{Z}_1)$ taken as positive, $|\gamma_1|$ is always less than unity.

Current and voltage scattering parameters

To define the generalized scattering parameter of a 1-port network we consider a generator of emf E_1 and internal impedance $\hat{Z}_1$, equal to the reference impedance, connected to the given network as shown in Fig. (1.6.1). We denote the current through and voltage across the network by I_1 and V_1 respectively.

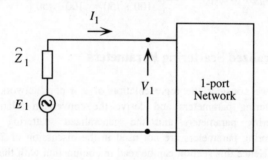

Fig. (1.6.1)

To proceed we first assume that the generator is disconnected from the network and connected to a load of impedance $\hat{Z}_1^*$, as shown in Fig. (1.6.2).

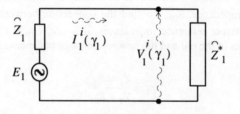

Fig. (1.6.2)

The current through and voltage across the impedance $\hat{Z}_1^*$ are defined as the incident current and incident voltage and are denoted by $I^i_1(\gamma_1)$ and $V^i_1(\gamma_1)$ respectively. The incident current $I^i_1(\gamma_1)$ and voltage $V^i_1(\gamma_1)$ in terms of the total current I_1 and voltage V_1 can be written as

$$I^i{}_1(\gamma_1) = \frac{E_1}{\hat{Z}_1 + \hat{Z}_1{}^*} = \frac{E_1}{2\hat{R}_1} = \frac{V_1 + \hat{Z}_1 I_1}{2\hat{R}_1} \tag{1.6.3}$$

and

$$V^i{}_1(\gamma_1) = \frac{\hat{Z}_1{}^* E_1}{\hat{Z}_1 + \hat{Z}_1{}^*} = \frac{\hat{Z}_1{}^* E_1}{2\hat{R}_1} = \frac{\hat{Z}_1{}^*(V_1 + \hat{Z}_1 I_1)}{2\hat{R}_1} \tag{1.6.4}$$

where

$$\hat{R}_1 = \mathrm{Re}(\hat{Z}_1).$$

The incident voltage and the incident current are related as

$$V^i{}_1(\gamma_1) = \hat{Z}_1{}^* I^i{}_1(\gamma_1). \tag{1.6.5}$$

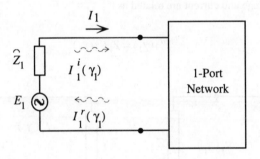

Fig. (1.6.3)

Next referring to Fig. (1.6.3), we define the reflected current $I^r{}_1(\gamma_1)$ by the expression

$$I^r{}_1(\gamma_1) = I^i{}_1(\gamma_1) - I_1$$

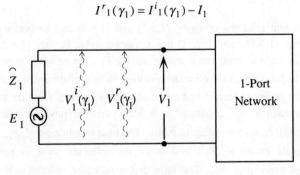

Fig. (1.6.4)

and similarly by referring to Fig. (1.6.4) we define the reflected voltage as

$$V^r{}_1(\gamma_1) = V_1 - V^i{}_1(\gamma_1).$$

In terms of the total currents and voltages,

$$I^r{}_1(\gamma_1) = \frac{V_1 + \hat{Z}_1 I_1}{2\hat{R}_1} - I_1 = \frac{V_1 - \hat{Z}_1^* I_1}{2\hat{R}_1} \tag{1.6.6}$$

and

$$V^r{}_1(\gamma_1) = V_1 - \frac{\hat{Z}_1^*(V_1 + \hat{Z}_1 I_1)}{2\hat{R}_1} = \frac{\hat{Z}_1(V_1 - \hat{Z}_1^* I_1)}{2\hat{R}_1}. \tag{1.6.7}$$

The reflected voltage and current are related as

$$V^r{}_1(\gamma_1) = \hat{Z}_1 I^r{}_1(\gamma_1). \tag{1.6.8}$$

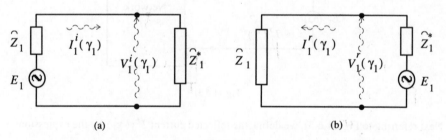

(a)   (b)

Fig. (1.6.5)

The voltage-current relations given by (1.6.5) and (1.6.8) can be better understood by referring to Fig. (1.6.5a) and Fig. (1.6.5b), respectively. In Fig. (1.6.5a), the source impedance is $\hat{Z}_1$ and the generator is terminated by $\hat{Z}_1^*$, the reflected current and voltage are both zero and the ratio of the incident voltage to that of the incident current is $\hat{Z}_1^*$. Similarly for the reflected current and reflected voltage, we can visualize a source having an internal impedance $\hat{Z}_1^*$ connected to a load which is now the source impedance $\hat{Z}_1$ (E_1 set to zero), as shown in Fig. (1.6.5b). The load termination of $\hat{Z}_1$, in this case, is again a conjugate match to $\hat{Z}_1^*$, and hence the reflected wave is not subsequently reflected at the impedance $\hat{Z}_1$. The ratio of the reflected voltage to that of reflected current is clearly $\hat{Z}_1$ as given by (1.6.8).

The scattering parameters are defined as the ratios of the reflected to that of incident current or voltage wave amplitudes. Hence the scattering parameter for the current can be written as

$$S_{11}{}^I(\gamma_1) = \frac{I^r{}_1(\gamma_1)}{I^i{}_1(\gamma_1)} = \frac{V_1 - \hat{Z}_1{}^* I_1}{V_1 + \hat{Z}_1 I_1} \tag{1.6.9}$$

and the scattering parameter for voltage as

$$S_{11}{}^V(\gamma_1) = \frac{V^r{}_1(\gamma_1)}{V^i{}_1(\gamma_1)} = \frac{\hat{Z}_1}{\hat{Z}_1{}^*} \frac{V_1 - \hat{Z}_1{}^* I_1}{V_1 + \hat{Z}_1 I_1}. \tag{1.6.10}$$

The wave amplitude scattering parameter

From (1.6.9) and (1.6.10), the scattering parameters for the voltage and current are clearly proportional and their ratio is the ratio of $\hat{Z}_1$ to $\hat{Z}_1{}^*$. Hence instead of considering two scattering parameters, we define a single scattering parameter as the ratio of the reflected wave amplitude to that of the incident wave amplitude as defined below.

We define the incident wave amplitude as

$$a_1(\gamma_1) = \sqrt{\hat{R}_1} I^i{}_1(\gamma_1) = \frac{\sqrt{\hat{R}_1}}{\hat{Z}_1{}^*} V^i{}_1(\gamma_1) \tag{1.6.11}$$

and the reflected wave amplitude as

$$b_1(\gamma_1) = \sqrt{\hat{R}_1} I^r{}_1(\gamma_1) = \frac{\sqrt{\hat{R}_1}}{\hat{Z}_1} V^r{}_1(\gamma_1) \tag{1.6.12}$$

and hence the wave amplitude scattering parameter as

$$S_{11}(\gamma_1) = \frac{b_1(\gamma_1)}{a_1(\gamma_1)} = \frac{I^r{}_1(\gamma_1)}{I^i{}_1(\gamma_1)} = \frac{\hat{Z}^*{}_1}{\hat{Z}_1} \frac{V^r{}_1(\gamma_1)}{V^i{}_1(\gamma_1)} = \frac{V_1 - \hat{Z}_1{}^* I_1}{V_1 + \hat{Z}_1 I_1}. \tag{1.6.13}$$

The new scattering parameter is related to that of the current and voltage scattering parameters by the expressions

$$S_{11}(\gamma_1) = \sqrt{\hat{R}_1} S_{11}(\gamma_1)^I \frac{1}{\sqrt{\hat{R}_1}} \tag{1.6.14}$$

and

$$S_{11}(\gamma_1) = \frac{\sqrt{\hat{R}_1}}{\hat{Z}_1} S_{11}(\gamma_1)^V \frac{\hat{Z}_1^*}{\sqrt{\hat{R}_1}}.$$ (1.6.15)

The terms $\sqrt{\hat{R}_1}$ and $1/\sqrt{\hat{R}_1}$ are included here in conformity with the corresponding expressions for multi-port networks.

The conversion of the wave amplitude scattering parameters of a 1-port network

As we have seen, the network scattering parameters can be defined and calculated for any reference impedance $\hat{Z}_1$. For measurement of the scattering parameters, however, a source with resistive impedance is used. We take this source resistance as our standard resistive reference and denote it by R_{01}.

The scattering parameter in a system of standard resistive reference $\hat{Z}_1 = R_{01}$ ($\gamma_1 = 0$) can be denoted as $S_{11}(0)$. For convenience, however, we use the notation S_{11} to stand for $S_{11}(0)$. We now consider the relation between a general scattering parameter $S_{11}(\gamma_1)$ and the measured scattering parameter S_{11}.

To find the conversion expression between the scattering parameters we note that the ratio

$$Z_{11} = \frac{V_1}{I_1}$$

is the impedance of the network, independent of the reference impedance. Hence the evaluation of Z_{11} can be used as an intermediary step for the conversion of the scattering parameters of different references. From (1.6.13)

$$S_{11}(\gamma_1) = \frac{V_1 - \hat{Z}_1^* I_1}{V_1 + \hat{Z}_1 I_1} = \frac{Z_{11} - \hat{Z}_1^*}{Z_{11} + \hat{Z}_1}$$ (1.6.16)

and hence

$$Z_{11} = \frac{S_{11}(\gamma_1)\hat{Z}_1 + \hat{Z}_1^*}{1 - S_{11}(\gamma_1)}$$ (1.6.17)

For the special case of $\hat{Z}_1 = R_{01}$,

$$Z_{11} = \frac{1 + S_{11}}{1 - S_{11}} R_{01}.$$ (1.6.18)

Eliminating Z_{11} between (1.6.17) and (1.6.18) and substituting $\hat{Z}_1$ in terms of γ_1, we have

$$S_{11}(\gamma_1) = \frac{1-\gamma_1}{1-\gamma_1^*} \frac{S_{11}-\gamma_1^*}{1-\gamma_1 S_{11}} \qquad (1.6.19)$$

which is the required conversion expression for the scattering parameters.

With $\gamma_1 = 0$, $S_{11}(\gamma_1) = S_{11}$, as expected. With $\gamma_1 = S_{11}^*$, from (1.6.2) and (1.6.18) $Z_{11}^* = Z_{IN}^* = \hat{Z}_1$ or Z_{11} is conjugate matched to $\hat{Z}_1$ and hence $S_{11}(\gamma_1) = 0$ as verified by (1.6.19).

Expressions for power in a system of reference coefficient $\gamma_1 = \Gamma_S$

When $\gamma_1 = \Gamma_S$ ($\hat{Z}_1 = Z_S$) the power delivered to the port or dissipated in the network P_{IN} is given by

$$P_{IN} = a_1(\Gamma_S) a_1^*(\Gamma_S) - b_1(\Gamma_S) b_1^*(\Gamma_S)$$

$$= a_1(\Gamma_S) a_1^*(\Gamma_S) [1 - |S_{11}(\Gamma_S)|^2] = e_S^2 [1 - |S_{11}(\Gamma_S)|^2] \qquad (1.6.20)$$

where $e_S = E_S/(2\sqrt{R_S})$ as defined before.

With $Z_{IN} = Z_S^*$, we have $S_{11}(\Gamma_S) = 0$ and the maximum available power from the source is given by the simple expression

$$P_{AVS} = e_S^2. \qquad (1.6.21)$$

We have defined the power gain G_T, as the ratio of the input power to available power. In terms of generalized scattering parameters, this also simplifies to

$$G_T = 1 - |S_{11}(\Gamma_S)|^2. \qquad (1.6.22)$$

Example (1.6.1)

Find the current, voltage and the wave amplitude scattering parameters for a 1-port network of impedance $Z_{11} = 75 + j100$ and for references (i) $\hat{Z}_1 = 50$ (ii) $\hat{Z}_1 = 25 - j50$ and (iii) $\hat{Z}_1 = 75 - j100$. Take the measured resistive reference as $R_{01} = 50$.

Solution

(I) For the case of $\hat{Z}_1 = \hat{Z}_1^* = R_{01} = 50$,

$$S_{11} = S_{11}^V = S_{11}^I = \frac{Z_{11} - R_{01}}{Z_{11} + R_{01}} = \frac{75 - j100 - 50}{75 - j100 + 50} = 0.512 + j0.390.$$

(ii) For the case $\hat{Z}_1 = 25 - j50$, using (1.6.1) we have,

$$\gamma_1 = (\hat{Z}_1 - R_{01})/(\hat{Z}_1 + R_{01}) = (25 - j50 - 50)/(25 - j50 + 50) = 0.077 - j0.615$$

and hence from (1.6.9) and with $V_1 = Z_{11}I_1$, we have

$$S_{11}(\gamma_1)^I = I^r{}_1(\gamma_1)/I^i{}_1(\gamma_1) = (Z_{11} - \hat{Z}_1^*)/(Z_{11} + \hat{Z}_1)$$

$$= (75 + j100 - 25 - j50)/(25 - j50 + 75 + j100) = 0.6 + j0.2$$

as the generalized current scattering parameter.

Similarly from (1.6.10) we find the generalized voltage scattering parameter as

$$S_{11}(\gamma_1)^V = V^r{}_1(\gamma_1)/V^i{}_1(\gamma_1) = [(Z_{11} - \hat{Z}_1^*)\hat{Z}_1/(Z_{11} + \hat{Z}_1)\hat{Z}_1^*]$$

$$= (0.6 + j0.2)(-0.6 - j0.8) = -0.2 - j0.6.$$

Finally from (1.6.14), we can write the generalized wave amplitude scattering parameter as

$$S_{11}(\gamma_1) = S_{11}(\gamma_1)^I = 0.6 + j0.2.$$

(iii) For the case of $Z_1 = 75 - j100$, using (1.6.1)

$$\gamma_1 = (\hat{Z}_1 - R_{01})/(\hat{Z}_1 + R_{01}) = (25 - j100)/(125 - j100) = 0.512 - j0.392.$$

The generalized current scattering parameter is again given by (1.6.9) as

$$S_{11}(\gamma_1)^I = I^r{}_1(\gamma_1)/I^i{}_1(\gamma_1) = (Z_{11} - \hat{Z}_1^*)/(Z_{11} + \hat{Z}_1)$$

$$= (75 + j100 - 75 - j100)/(75 - j100 + 75 + j100) = 0.$$

Similarly from (1.6.10) and (1.6.14) the voltage and wave amplitude scattering parameters are both zero. This is expected as the input impedance of the network is a conjugate of the reference impedance.

Example (1.6.2)

For the cases (ii) and (iii) of the above example show that the relation

$$S_{11}(\gamma_1) = \frac{1-\gamma_1}{1-\gamma_1^*} \frac{S_{11} - \gamma_1^*}{1 - \gamma_1 S_{11}}$$

given by (1.6.19) is satisfied.

Solution

For this case (ii)

$$S_{11} = 0.512 + j0.390, \gamma_1 = 0.077 - j0.615$$

and after substitution

$$S_{11}(\gamma_1) = \frac{1 - 0.077 + j0.615}{1 - 0.077 - j0.615} \frac{0.512 + j.390 - 0.077 - j0.615}{1 - (0.077 - j0.615)(0.512 + j0.390)} = 0.6 + j0.2$$

which is the value found in the previous example.

For the case (iii)

$$S_{11} = 0.512 + j0.390, \gamma_1 = 0.512 - j0.390,$$

and hence $S_{11} - \gamma_1^* = 0$, or $S_{11}(\gamma_1) = 0$, as found before.

Example (1.6.3)

For Example (1.5.1), calculate the power dissipated in the load using the generalized scattering parameters with the reference impedance $\widehat{Z}_1 = Z_S$.

Solution

The scattering parameter $S_{11}(\Gamma_S)$ for $\widehat{Z}_1 = Z_S$ was found in Example (1.6.1) as $S_{11}(\Gamma_S) = 0.6 + j0.2$ and hence $\left|S_{11}(\Gamma_S)\right|^2 = 0.4$, giving

$$P_{IN} = e_S^2[1 - \left|S_{11}(\Gamma_S)\right|^2] = 0.25(1 - 0.4) = 0.15$$

which was found previously in Example (1.5.1).

Chapter 2

The Scattering Parameters
of a 2-port Network

The 2-port network is the most commonly used network in practical applications. This justifies its discussion in a separate chapter. In addition, a better understanding of the general scattering parameter analysis can be gained by a detailed study of the 2-port network, prior to the consideration of an N-port network. In this Chapter we define the wave amplitudes and the scattering parameters of a 2-port network, in a system of resistive reference impedances, and establish the relations between these parameters. The relation between the 2-port scattering parameters and other circuit representation such as impedance and admittance parameters is also given. A so called 'Flow Graph' in conjunction with 'Mason's rule' simplifies the derivations of the wave amplitudes in a circuit of given scattering parameters.

2.1 Scattering Matrix of a 2-port Network

As we have discussed in Chapter 1, the scattering parameters are defined for a set of reference impedances. In the case of a 2-port network we require two reference impedances that can be denoted as $\hat{Z}_1$ and $\hat{Z}_2$, for port 1 and port 2 respectively. In this chapter we assume that $\hat{Z}_1 = R_{01}$ and $\hat{Z}_2 = R_{02}$ are both real and equal to R_0. The case of reference impedances not being purely resistive or not equal is discussed in Chapter 6.

Current and voltage scattering matrices

As the concepts of current and voltage are more familiar, we define the incident and reflected currents and voltages and the related scattering parameters, before a discussion of the wave amplitude scattering parameters.

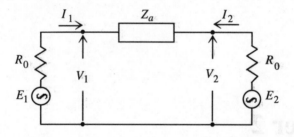

Fig. (2.1.1)

To define the scattering parameters of a 2-port network in a system of reference impedances R_0, we assume a network consisting of two generators with assumed internal impedances equal to R_0 as shown in Fig. (2.1.1). The generator emf of port 1 can be denoted by E_1, the current flowing into port 1 by I_1 and the voltage across the same port by V_1.

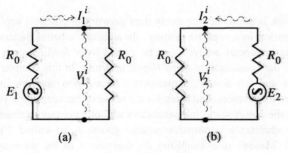

(a)   (b)

Fig. (2.1.2)

The parameters relating to port 2 can similarly be defined, but with the subscripts changing from 1 to 2.

We define the incident currents $\mathbf{I}^i$ and voltages $\mathbf{V}^i$ as the currents through and voltages across the two ports, when the 2-port network is disconnected and each generator is terminated by its reference impedance, as in Fig. (2.1.2).

Hence by referring to the above figure, we can write

$$\mathbf{I}^i = \begin{bmatrix} I_1^i \\ I_2^i \end{bmatrix} = \begin{bmatrix} \dfrac{E_1}{2R_0} \\ \dfrac{E_2}{2R_0} \end{bmatrix} = \begin{bmatrix} \dfrac{V_1 + R_0 I_1}{2R_0} \\ \dfrac{V_2 + R_0 I_2}{2R_0} \end{bmatrix}. \tag{2.1.1}$$

Similarly

$$\mathbf{V}^i = \begin{bmatrix} V_1^i \\ V_2^i \end{bmatrix} = \begin{bmatrix} \dfrac{E_1}{2} \\ \dfrac{E_2}{2} \end{bmatrix} = \begin{bmatrix} \dfrac{V_1 + R_0 I_1}{2} \\ \dfrac{V_2 + R_0 I_2}{2} \end{bmatrix}. \tag{2.1.2}$$

From (2.1.1) and (2.1.2)

$$\mathbf{V}^i = \begin{bmatrix} R_0 & 0 \\ 0 & R_0 \end{bmatrix} \mathbf{I}^i. \tag{2.1.3}$$

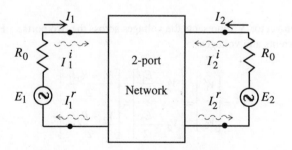

Fig. (2.1.3a)

Now replacing the network as in Fig. (2.1.3a), we define the reflected current from the relation

$$\mathbf{I}^r = \mathbf{I}^i - \mathbf{I} \tag{2.1.4}$$

where the column vector $\mathbf{I}$ represents the total currents in both ports.

Hence in terms of the total currents and voltages

$$\mathbf{I}^r = \begin{bmatrix} I_1^r \\ I_2^r \end{bmatrix} = \begin{bmatrix} \dfrac{E_1}{2R_0} - I_1 \\ \dfrac{E_2}{2R_0} - I_2 \end{bmatrix} = \begin{bmatrix} \dfrac{V_1 - R_0 I_1}{2R_0} \\ \dfrac{V_2 - R_0 I_2}{2R_0} \end{bmatrix}. \tag{2.1.5}$$

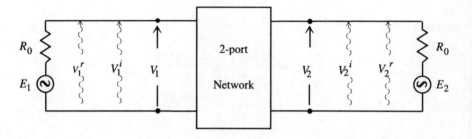

Fig. (2.1.3b)

Similarly referring to Fig. (2.1.3b) we define the reflected voltage by the relation

$$\mathbf{V}^r = \mathbf{V} - \mathbf{V}^i \tag{2.1.6}$$

where the column vector $\mathbf{V}$ represents the voltages across the two ports.   Hence again in terms of total currents and voltages

$$\mathbf{V}^r = \begin{bmatrix} V_1^r \\ V_2^r \end{bmatrix} = \begin{bmatrix} V_1 - \dfrac{E_1}{2} \\ V_2 - \dfrac{E_2}{2} \end{bmatrix} = \begin{bmatrix} \dfrac{V_1 - R_0 I_1}{2} \\ \dfrac{V_2 - R_0 I_2}{2} \end{bmatrix}. \tag{2.1.7}$$

From (2.1.5) and (2.1.7)

$$\mathbf{V}^r = \begin{bmatrix} R_0 & 0 \\ 0 & R_0 \end{bmatrix} \mathbf{I}^r. \tag{2.1.8}$$

The relation between $\mathbf{I}^r$ and $\mathbf{I}^i$ can be written as

$$\mathbf{I}^r = \mathbf{S}^I \mathbf{I}^i \qquad (2.1.9)$$

where

$$\mathbf{S}^I = \begin{bmatrix} S_{11}^I & S_{12}^I \\ S_{21}^I & S_{22}^I \end{bmatrix} \qquad (2.1.10)$$

defines the current scattering matrix of the 2-port network in a system of reference impedances R_0. Similarly

$$\mathbf{V}^r = \mathbf{S}^V \mathbf{V}^i \qquad (2.1.11)$$

where

$$\mathbf{S}^V = \begin{bmatrix} S_{11}^V & S_{12}^V \\ S_{21}^V & S_{22}^V \end{bmatrix} \qquad (2.1.12)$$

defines the voltage scattering matrix.

If the generator impedances are not equal to the reference impedances, the incident and reflected currents and voltages are modified. The defined scattering parameters, however, remain unchanged. This is further discussed in Section (2.5).

Wave amplitude scattering matrix

As the incident and reflected currents and voltages are related by expressions (2.1.3) and (2.1.8), it is not necessary to consider both the incident and reflected currents and voltages, and hence the current and voltage scattering parameters. Instead, as in the case of 1-port networks, we define a set of incident and reflected wave amplitudes and hence a single wave amplitude scattering matrix. As we shall see the relevant power expressions are also more conveniently expressed in terms of these defined wave amplitudes.

The incident wave amplitudes are related to the incident currents and voltages and are defined as

$$\mathbf{a} = \begin{bmatrix} a_1 \\ a_2 \end{bmatrix} = \begin{bmatrix} \sqrt{R_0} & 0 \\ 0 & \sqrt{R_0} \end{bmatrix} \mathbf{I}^i = \begin{bmatrix} \sqrt{G_0} & 0 \\ 0 & \sqrt{G_0} \end{bmatrix} \mathbf{V}^i \qquad (2.1.13)$$

where $G_0 = 1/R_0$.

Similarly the reflected wave amplitudes are defined as

$$\mathbf{b} = \begin{bmatrix} b_1 \\ b_2 \end{bmatrix} = \begin{bmatrix} \sqrt{R_0} & 0 \\ 0 & \sqrt{R_0} \end{bmatrix} \mathbf{I}^r = \begin{bmatrix} \sqrt{G_0} & 0 \\ 0 & \sqrt{G_0} \end{bmatrix} \mathbf{V}^r. \qquad (2.1.14)$$

In terms of the total currents and voltages the incident and reflected wave amplitudes $\mathbf{a}$ and $\mathbf{b}$ are given by the expressions

$$\mathbf{a} = \begin{bmatrix} \dfrac{V_1 + R_0 I_1}{2\sqrt{R_0}} \\[3mm] \dfrac{V_2 + R_0 I_2}{2\sqrt{R_0}} \end{bmatrix} \qquad (2.1.15)$$

and

$$\mathbf{b} = \begin{bmatrix} \dfrac{V_1 - R_0 I_1}{2\sqrt{R_0}} \\[3mm] \dfrac{V_2 - R_0 I_2}{2\sqrt{R_0}} \end{bmatrix}. \qquad (2.1.16)$$

The matrix relation

$$\mathbf{b} = \mathbf{S}\mathbf{a} \qquad (2.1.17)$$

where

$$\mathbf{S} = \begin{bmatrix} S_{11} & S_{12} \\ S_{21} & S_{22} \end{bmatrix} \qquad (2.1.18)$$

defines the wave amplitude scattering matrix of the network. S_{11}, S_{12}, S_{21} and S_{22} are known as the 2-port network wave amplitude scattering parameters, or simply as the scattering parameters of the 2-port network.

From equations (2.1.13) and (2.1.14) we can write $\mathbf{I}^i$, $\mathbf{V}^i$, $\mathbf{I}^r$ and $\mathbf{V}^r$ in terms of $\mathbf{a}$ and $\mathbf{b}$ by the following expressions

$$\mathbf{I}^i = \begin{bmatrix} \sqrt{G_0} & 0 \\ 0 & \sqrt{G_0} \end{bmatrix} \mathbf{a}, \quad \mathbf{V}^i = \begin{bmatrix} \sqrt{R_0} & 0 \\ 0 & \sqrt{R_0} \end{bmatrix} \mathbf{a}$$

and

$$\mathbf{I}^r = \begin{bmatrix} \sqrt{G_0} & 0 \\ 0 & \sqrt{G_0} \end{bmatrix} \mathbf{b}, \quad \mathbf{V}^r = \begin{bmatrix} \sqrt{R_0} & 0 \\ 0 & \sqrt{R_0} \end{bmatrix} \mathbf{b}.$$

Hence it can easily be verified that

$$\mathbf{S} = \mathbf{S}^I = \mathbf{S}^V.$$

We should realise that the above relation is only true in this particular situation where the reference impedances are both assumed to be resistive.

Example (2.1.1)

In a system of reference impedances R_0, find the scattering matrix of a 2-port network consisting of a series impedance Z_a, as shown

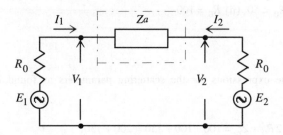

Solution

The scattering matrix of the network can be found by a direct comparison of the wave amplitudes **a** and **b**. Referring to the above circuit diagram and from (2.1.15) and (2.1.16) we can readily write **a** and **b** as

$$\mathbf{a} = \begin{bmatrix} \dfrac{E_1}{2\sqrt{R_0}} \\[2ex] \dfrac{E_2}{2\sqrt{R_0}} \end{bmatrix}, \quad \mathbf{b} = \begin{bmatrix} \dfrac{Z_a}{2\sqrt{R_0}\,Z} E_1 + \dfrac{\sqrt{R_0}}{Z} E_2 \\[2ex] \dfrac{\sqrt{R_0}}{Z} E_1 + \dfrac{Z_a}{2\sqrt{R_0}\,Z} E_2 \end{bmatrix}$$

where we have made the following substitutions:

$$V_1 = E_1 - R_0 I_1, \qquad I_1 = \frac{(E_1 - E_2)}{Z},$$

$$V_2 = E_2 - R_0 I_2, \qquad I_2 = \frac{(E_2 - E_1)}{Z}$$

where $Z = 2R_0 + Z_a$.

Comparing the expressions for **b** and **a**, the scattering parameters can be found as

$$S_{11} = \frac{Z_a}{Z}, \qquad S_{12} = \frac{2R_0}{Z},$$

$$S_{21} = \frac{2R_0}{Z}, \qquad S_{22} = \frac{Z_a}{Z}.$$

Example (2.1.2)

Find the scattering parameters of a series impedance $Z_a = 100 + j50$ for reference impedances (i) $R_0 = 50$, (ii) $R_0 = 100$

Solution

Here we use the expressions for the scattering parameters as found in the previous example.

(i) $Z = 2R_0 + Z_a = 100 + 100 + j50 = 200 + j50,$

and after substitution

$$S_{11} = S_{22} = 0.529 + j0.118, \quad S_{12} = S_{21} = 0.471 - j0.118.$$

(ii) $Z = 2R_0 + Z_a = 200 + 100 + j50 = 300 + j50,$

and after substitution

$$S_{11} = S_{22} = 0.351 + j0.108, \quad S_{12} = S_{21} = 0.649 - j0.108.$$

From the above examples we can see that the scattering parameters in two systems with different reference impedances are different.

Example (2.1.3)

In a system of reference impedances R_0, find the scattering matrix of a 2-port network consisting of a shunt admittance Y_b.

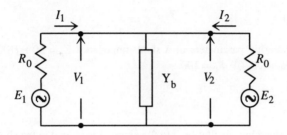

Solution

Referring to the above circuit diagram and from (2.1.15) and (2.1.16) and by substituting

$$V_1 = E_1 - R_0 I_1, \qquad I_1 = \frac{(Y - G_0)E_1}{R_0 Y} - \frac{G_0 E_2}{R_0 Y},$$

$$V_2 = E_2 - R_0 I_2, \qquad I_2 = \frac{(Y - G_0)E_2}{R_0 Y} - \frac{G_0 E_1}{R_0 Y}$$

where $Y = 2G_0 + Y_b$ and $G_0 = 1/R_0$, we have

$$\mathbf{a} = \begin{bmatrix} \dfrac{E_1}{2\sqrt{R_0}} \\[3mm] \dfrac{E_2}{2\sqrt{R_0}} \end{bmatrix}$$

and

$$\mathbf{b} = \begin{bmatrix} -\dfrac{Y_b}{2\sqrt{R_0}\,Y} E_1 + \dfrac{G_0}{\sqrt{R_0}\,Y} E_2 \\[4mm] \dfrac{G_0}{\sqrt{R_0}\,Y} E_1 - \dfrac{Y_b}{2\sqrt{R_0}\,Y} E_2 \end{bmatrix}.$$

From the above expressions for **a** and **b**, the scattering parameters are given as

$$S_{11} = \frac{-Y_b}{Y}, \quad S_{12} = \frac{2G_0}{Y},$$

$$S_{21} = \frac{2G_0}{Y}, \quad S_{22} = \frac{-Y_b}{Y}.$$

Example (2.1.4)

Calculate the scattering parameters of a shunt impedance $Z_b = 50 + j100$ for reference impedances (i) $R_0 = 50$, (ii) $R_0 = 100$.

Solution

The problem is solved by substitution into the expressions found in the above example.

(i) $Y = (0.004 - j0.008) + 0.020 + 0.020 = 0.044 - j0.008, \quad G_0 = 0.020$, and hence

$$S_{11} = S_{22} = -0.120 + j0.160, \quad S_{12} = S_{21} = 0.880 + j0.160.$$

(ii) $Y = (0.004 - j0.008) + 0.010 + 0.010 = 0.024 - j0.008, \quad G_0 = 0.010$, and hence

$$S_{11} = S_{22} = -0.250 + j0.250, \quad S_{21} = S_{12} = 0.750 + j0.250.$$

2.2 The Physical Interpretation and the Measurement of the Scattering Parameters

For a 2-port network, the scattering matrix $\mathbf{S}$ relates the reflected wave amplitude matrix $\mathbf{b}$ to the incident wave amplitude matrix $\mathbf{a}$ by the matrix relation

$$\mathbf{b} = \mathbf{S}\,\mathbf{a}$$

or in expanded form as

$$b_1 = S_{11}a_1 + S_{12}a_2,$$

$$b_2 = S_{21}a_1 + S_{22}a_2.$$

The scattering parameters S_{11}, S_{12}, S_{21} and S_{22} can hence be defined as

$$S_{11} = \left. \frac{b_1}{a_1} \right|_{a_2 = 0},$$ (2.2.1a)

$$S_{12} = \left. \frac{b_1}{a_2} \right|_{a_1 = 0},$$ (2.2.1b)

$$S_{21} = \left. \frac{b_2}{a_1} \right|_{a_2 = 0},$$ (2.2.1c)

$$S_{22} = \left. \frac{b_2}{a_2} \right|_{a_1 = 0}.$$ (2.2.1d)

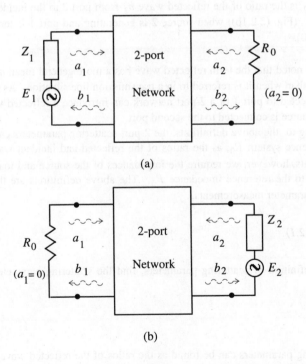

(a)

(b)

Fig. (2.2.1)

From the definition of the incident wave as given by (2.1.15), $a_1 = 0$ when $V_1 + R_0 I_1 = 0$, and hence when $Z_1 = R_0$ and $E_1 = 0$. Similarly $a_2 = 0$ when $V_2 + R_0 I_2 = 0$ and hence when $Z_2 = R_0$ and $E_2 = 0$. Therefore, referring to Fig. (2.2.1), the scattering parameters can be defined in words as

(1) S_{11} is the ratio of the reflected wave b_1 from port 1 to the incident wave a_1 on port 1 (Fig. (2.2.1a)) when source 1 is generating and port 2 is terminated by R_0 ($a_2 = 0$).

(2) S_{12} is the ratio of the reflected wave b_1 from port 1 to the incident wave a_2 on port 2 (Fig. (2.2.1b)) when source 2 is generating and port 1 is terminated by R_0 ($a_1 = 0$).

(3) S_{21} is the ratio of the reflected wave b_2 from port 2 to the incident wave a_1 on port 1 (Fig. (2.2.1a)) when source 1 is generating and port 2 is terminated by R_0 ($a_2 = 0$).

(4) S_{22} is the ratio of the reflected wave b_2 from port 2 to the incident wave a_2 on port 2 (Fig. (2.2.1b)) when source 2 is generating and port 1 is terminated by R_0 ($a_1 = 0$).

It should be noted that the term reflected wave has a more general meaning here in these definitions, than when it is referred to in a transmission line situation. As defined here, a source connected to port 1 of a 2-port network can introduce a reflected wave in port 2, even if no source is connected to the second port.

According to the above definitions, the 2-port scattering parameters can be measured in any reference system R_0, as the ratios of the reflected and incident waves. For such measurements, however, we require the impedances of the source and load terminations to be equal to the reference impedance R_0. The above definitions are the basis of the scattering parameter measurements.

Example (2.2.1)

From the definition of scattering parameters, find the scattering parameters of problem (2.1.1)

Solution

The scattering parameters can be found as the ratios of the reflected wave amplitudes to that of the incident wave amplitudes. For this particular problem, however, it is more convenient to consider the incident and reflected currents. The situation is depicted in figure (a).

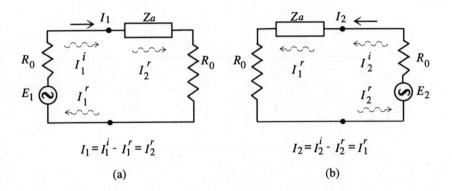

$$I_1 = I_1^i - I_1^r = I_2^r \qquad\qquad I_2 = I_2^i - I_2^r = I_1^r$$

(a) (b)

The scattering parameter S_{11} can be found by definition (1) and the fact that $\mathbf{S} = \mathbf{S}^t$, as the ratio of the reflected current to that of the incident current for port 1, when port 2 is terminated by the reference impedance R_0. The incident and reflected currents are readily seen to be $I_1^i = E_1/(2R_0)$ and $I_1^r = Z_a E_1/(2R_0 Z)$ and hence we have

$$S_{11} = \frac{b}{a} = \frac{I_1^r}{I_1^i} = \frac{Z_a}{Z}$$

where again $Z = 2R_0 + Z_a$.

To find S_{21} we note that in the case of the network consisting of a series impedance $I_2^r = I_1^i - I_1^r = I_1$ and hence $I_2^r = E_1/Z$, or from definition (3)

$$S_{21} = \frac{2R_0}{Z}.$$

Similarly by referring to figure (b) and definitions (2) and (4), we can write S_{12}, and S_{22} as

$$S_{12} = \frac{2R_0}{Z}, \qquad S_{22} = \frac{Z_a}{Z}$$

as in problem (2.1.1).

Example (2.2.2)

From the definition of scattering parameters, find the scattering parameters of problem (2.1.3).

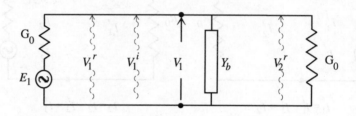

Solution

The procedure in this case is similar to that of the last problem. For this case, however, if port 1 is generating, the reflected voltage from port 2 is equal to the sum of the incident and reflected voltages related to port 1. Similarly if port 2 is generating, the reflected voltage from port 1 is equal to the sum of the incident and the reflected voltage related to port 2. Hence it is appropriate to find the scattering parameters by considering the incident and reflected voltages.

Referring to the above figure we can write $V_1 = V_1^i + V_1^r = V_2^r$ and easily verify that

$$S_{11} = \frac{1/(Y_b + G_0) - R_0}{R_0 + 1/(Y_b + G_0)} = \frac{-Y_b}{Y}$$

and

$$S_{21} = 1 + \frac{2G_0 - Y}{Y} = \frac{2G_0}{Y}$$

where $Y = 2G_0 + Y_b$. In a similar way we can find the two remaining scattering parameters as

$$S_{12} = \frac{2G_0}{Y}, \qquad S_{22} = \frac{-Y_b}{Y}.$$

2.3 Derivation of the Scattering Parameters from the Impedance and Admittance Parameters

The scattering parameters of a 2-port network are closely related to the other network representations, such as the impedance and admittance parameters. In this section we derive the conversion relations between these different representations.

The impedance parameters

The impedance parameters for a 2-port network can be defined by a matrix relation between the voltages and currents given in the form of

$$
\begin{bmatrix} V_1 \\ V_2 \end{bmatrix} = \begin{bmatrix} Z_{11} & Z_{12} \\ Z_{21} & Z_{22} \end{bmatrix} \begin{bmatrix} I_1 \\ I_2 \end{bmatrix}
$$

(2.3.1)

where V_1 and V_2 are the voltages across port 1 and port 2 respectively. Similarly the currents flowing into port 1 and port 2 are denoted by I_1 and I_2 respectively.

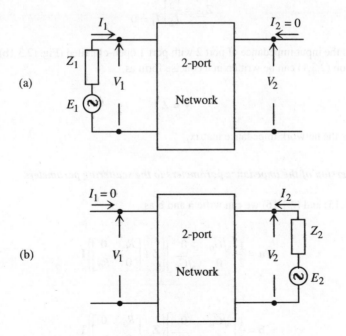

Fig. (2.3.1)

With this definition it is clear that

$$Z_{11} = \frac{V_1}{I_1}\bigg|_{I_2 = 0}$$

is the input impedance of port 1 with port 2 open-circuited (Fig. (2.3.1a)),

$$Z_{12} = \frac{V_1}{I_2}\bigg|_{I_1 = 0}$$

is the reverse transfer impedance with port 1 open-circuited (Fig. (2.3.1b)),

$$Z_{21} = \frac{V_2}{I_1}\bigg|_{I_2 = 0}$$

represents the forward transfer impedance with port 2 open-circuited (Fig. (2.3.1a)) and

$$Z_{22} = \frac{V_2}{I_2}\bigg|_{I_1 = 0}$$

represents the input impedance of port 2 with port 1 open-circuited (Fig. (2.3.1b)).
    Equation (2.3.1) can be written in a concise form as

$$\mathbf{V} = \mathbf{Z}\,\mathbf{I} \tag{2.3.2}$$

where $\mathbf{Z}$ is the network impedance matrix.

The conversion of the impedance parameters to the scattering parameters

From (2.1.15) and (2.1.16) we can write $\mathbf{a}$ and $\mathbf{b}$ as

$$\mathbf{a} = \frac{1}{2}\begin{bmatrix} \sqrt{G_0} & 0 \\ 0 & \sqrt{G_0} \end{bmatrix}\left\{\mathbf{Z} + \begin{bmatrix} R_0 & 0 \\ 0 & R_0 \end{bmatrix}\right\}\mathbf{I} \tag{2.3.3}$$

and

$$\mathbf{b} = \frac{1}{2}\begin{bmatrix} \sqrt{G_0} & 0 \\ 0 & \sqrt{G_0} \end{bmatrix}\left\{\mathbf{Z} - \begin{bmatrix} R_0 & 0 \\ 0 & R_0 \end{bmatrix}\right\}\mathbf{I} \tag{2.3.4}$$

and hence from the relation $\mathbf{b} = \mathbf{Sa}$ or $\mathbf{S} = \mathbf{ba}^{-1}$, we have

$$\mathbf{S} = \begin{bmatrix} \sqrt{G_0} & 0 \\ 0 & \sqrt{G_0} \end{bmatrix} \left\{ \mathbf{Z} - \begin{bmatrix} R_0 & 0 \\ 0 & R_0 \end{bmatrix} \right\} \left\{ \mathbf{Z} + \begin{bmatrix} R_0 & 0 \\ 0 & R_0 \end{bmatrix} \right\}^{-1} \begin{bmatrix} \sqrt{R_0} & 0 \\ 0 & \sqrt{R_0} \end{bmatrix}$$

$$= \left\{ \mathbf{Z} - \begin{bmatrix} R_0 & 0 \\ 0 & R_0 \end{bmatrix} \right\} \left\{ \mathbf{Z} + \begin{bmatrix} R_0 & 0 \\ 0 & R_0 \end{bmatrix} \right\}^{-1} \tag{2.3.5}$$

relating the scattering parameters to the impedance parameters. Substituting for $\mathbf{Z}$ and expanding we have

$$\mathbf{S} = \frac{1}{\Sigma_{ZS}} \begin{bmatrix} (Z_{11} - R_0) & Z_{12} \\ Z_{21} & (Z_{22} - R_0) \end{bmatrix} \begin{bmatrix} (Z_{22} + R_0) & -Z_{12} \\ -Z_{21} & (Z_{11} + R_0) \end{bmatrix}$$

where

$$\Sigma_{ZS} = R_0 R_0 + R_0 Z_{11} + R_0 Z_{22} + \Delta_Z \tag{2.3.6}$$

and

$$\Delta_Z = Z_{11} Z_{22} - Z_{12} Z_{21}.$$

After matrix multiplication, the scattering parameters are found in terms of the impedance parameters as

$$S_{11} = \frac{1}{\Sigma_{ZS}} (R_0 Z_{11} - R_0 Z_{22} - R_0 R_0 + \Delta_Z), \tag{2.3.7a}$$

$$S_{12} = \frac{2}{\Sigma_{ZS}} R_0 Z_{12}, \tag{2.3.7b}$$

$$S_{21} = \frac{2}{\Sigma_{ZS}} R_0 Z_{21}, \tag{2.3.7c}$$

$$S_{22} = \frac{1}{\Sigma_{ZS}} (R_0 Z_{22} - R_0 Z_{11} - R_0 R_0 + \Delta_Z). \tag{2.3.7d}$$

The Conversion of the scattering parameters to the impedance parameters

To find the impedance parameters in terms of the scattering parameters, from expression (2.3.3) and (2.3.4) for **a** and **b** and **b** = **Sa** we can write

$$
\begin{bmatrix} \sqrt{G_0} & 0 \\ 0 & \sqrt{G_0} \end{bmatrix} \left\{ \mathbf{Z} - \begin{bmatrix} R_0 & 0 \\ 0 & R_0 \end{bmatrix} \right\} = \mathbf{S} \begin{bmatrix} \sqrt{G_0} & 0 \\ 0 & \sqrt{G_0} \end{bmatrix} \left\{ \mathbf{Z} + \begin{bmatrix} R_0 & 0 \\ 0 & R_0 \end{bmatrix} \right\}.
$$

With some rearrangements we can find **Z** as

$$
\mathbf{Z} = \begin{bmatrix} \sqrt{R_0} & 0 \\ 0 & \sqrt{R_0} \end{bmatrix} \{\mathbf{U} - \mathbf{S}\}^{-1} \left\{ \begin{bmatrix} \sqrt{G_0} & 0 \\ 0 & \sqrt{G_0} \end{bmatrix} \begin{bmatrix} R_0 & 0 \\ 0 & R_0 \end{bmatrix} + \mathbf{S} \begin{bmatrix} \sqrt{G_0} & 0 \\ 0 & \sqrt{G_0} \end{bmatrix} \begin{bmatrix} R_0 & 0 \\ 0 & R_0 \end{bmatrix} \right\}
$$

$$
= \{\mathbf{U} - \mathbf{S}\}^{-1} \{\mathbf{U} + \mathbf{S}\} \begin{bmatrix} R_0 & 0 \\ 0 & R_0 \end{bmatrix}
$$

where **U** is a diagonal matrix of diagonal elements unity (identity matrix).
By matrix inversion and multiplication we find the impedance parameters in terms of the scattering parameters as

$$
Z_{11} = \frac{R_0}{\Sigma_{SZ}} (1 + S_{11} - S_{22} - \Delta_S), \tag{2.3.8a}
$$

$$
Z_{12} = \frac{2}{\Sigma_{SZ}} R_0 S_{12}, \tag{2.3.8b}
$$

$$
Z_{21} = \frac{2}{\Sigma_{SZ}} R_0 S_{21}, \tag{2.3.8c}
$$

$$
Z_{22} = \frac{R_0}{\Sigma_{SZ}} (1 + S_{22} - S_{11} - \Delta_S) \tag{2.3.8d}
$$

where

$$
\Sigma_{SZ} = 1 - S_{11} - S_{22} + \Delta_S \tag{2.3.9}
$$

and

$$
\Delta_S = S_{11}S_{22} - S_{12}S_{21}.
$$

The admittance parameters

The admittance parameters of a 2-port network can be defined as the elements of a matrix relating the currents and voltages of the network in the form of

$$\begin{bmatrix} I_1 \\ I_2 \end{bmatrix} = \begin{bmatrix} Y_{11} & Y_{12} \\ Y_{21} & Y_{22} \end{bmatrix} \begin{bmatrix} V_1 \\ V_2 \end{bmatrix}. \tag{2.3.10}$$

From (2.3.10) it is clear that

$$Y_{11} = \frac{I_1}{V_1}\bigg|_{V_2 = 0}$$

is the input admittance of port 1 with port 2 short-circuited (Fig. (2.3.2a)),

$$Y_{12} = \frac{I_1}{V_2}\bigg|_{V_1 = 0}$$

represents the reverse transfer admittance with port 1 short-circuited(Fig.(2.3.2b)),

$$Y_{21} = \frac{I_2}{V_1}\bigg|_{V_2 = 0}$$

represents the forward transfer admittance with port 2 short-circuited(Fig.(2.3.2a)) and

$$Y_{22} = \frac{I_2}{V_2}\bigg|_{V_1 = 0}$$

is the input admittance of port-2 with port 1 short-circuited(Fig. (2.3.2b)).

The matrix equation (2.3.10) can be written as

$$\mathbf{I} = \mathbf{Y}\,\mathbf{V}. \tag{2.3.11}$$

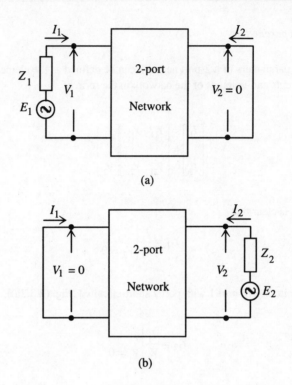

(a)

(b)

Fig. (2.3.2)

The conversion of the admittance parameters to the scattering parameters

From (2.1.15), (2.1.16) and (2.3.10)

$$\mathbf{a} = \frac{1}{2}\begin{bmatrix} \sqrt{G_0} & 0 \\ 0 & \sqrt{G_0} \end{bmatrix} \begin{bmatrix} R_0 & 0 \\ 0 & R_0 \end{bmatrix} \left\{ \begin{bmatrix} G_0 & 0 \\ 0 & G_0 \end{bmatrix} + \mathbf{Y} \right\} \mathbf{V} \tag{2.3.12}$$

and

$$\mathbf{b} = \frac{1}{2}\begin{bmatrix} \sqrt{G_0} & 0 \\ 0 & \sqrt{G_0} \end{bmatrix} \begin{bmatrix} R_0 & 0 \\ 0 & R_0 \end{bmatrix} \left\{ \begin{bmatrix} G_0 & 0 \\ 0 & G_0 \end{bmatrix} - \mathbf{Y} \right\} \mathbf{V}. \tag{2.3.13}$$

Hence with

$$\mathbf{b} = \mathbf{Sa},$$

we have

$$S = \begin{bmatrix} \sqrt{G_0} & 0 \\ 0 & \sqrt{G_0} \end{bmatrix} \left\{ \begin{bmatrix} G_0 & 0 \\ 0 & G_0 \end{bmatrix} - Y \right\} \left\{ \begin{bmatrix} G_0 & 0 \\ 0 & G_0 \end{bmatrix} + Y \right\}^{-1} \begin{bmatrix} \sqrt{R_0} & 0 \\ 0 & \sqrt{R_0} \end{bmatrix}$$

$$= \left\{ \begin{bmatrix} G_0 & 0 \\ 0 & G_0 \end{bmatrix} - Y \right\} \left\{ \begin{bmatrix} G_0 & 0 \\ 0 & G_0 \end{bmatrix} + Y \right\}^{-1}. \tag{2.3.14}$$

Substitution for Y in terms of the admittance parameters, we have

$$S = \frac{1}{\Sigma_{YS}} \begin{bmatrix} (G_0 - Y_{11}) & -Y_{12} \\ -Y_{21} & (G_0 - Y_{22}) \end{bmatrix} \begin{bmatrix} (G_0 + Y_{22}) & -Y_{12} \\ -Y_{21} & (G_0 + Y_{11}) \end{bmatrix}$$

where

$$\Sigma_{YS} = G_0 G_0 + G_0 Y_{11} + G_0 Y_{22} + \Delta_Y \tag{2.3.15}$$

and

$$\Delta_Y = Y_{11} Y_{22} - Y_{12} Y_{21}.$$

Hence the scattering parameters in terms of admittance parameters are

$$S_{11} = \frac{1}{\Sigma_{YS}} (G_0 G_0 - Y_{11} G_0 + G_0 Y_{22} - \Delta_Y), \tag{2.3.16a}$$

$$S_{12} = -\frac{2}{\Sigma_{YS}} G_0 Y_{12}, \tag{2.3.16b}$$

$$S_{21} = -\frac{2}{\Sigma_{YS}} G_0 Y_{21}, \tag{2.3.16c}$$

$$S_{22} = \frac{1}{\Sigma_{YS}} (G_0 G_0 - Y_{22} G_0 + G_0 Y_{11} - \Delta_Y). \tag{2.3.16d}$$

The conversion of the scattering parameters to the admittance parameters

To find the admittance parameters in terms of the scattering parameters, we write (2.3.14) as

$$\left\{ \begin{bmatrix} G_0 & 0 \\ 0 & G_0 \end{bmatrix} - \mathbf{Y} \right\} = \mathbf{S} \left\{ \begin{bmatrix} G_0 & 0 \\ 0 & G_0 \end{bmatrix} + \mathbf{Y} \right\}.$$

After rearrangement we can find $\mathbf{Y}$ as

$$\mathbf{Y} = \{\mathbf{U} + \mathbf{S}\}^{-1} \{\mathbf{U} - \mathbf{S}\} \begin{bmatrix} G_0 & 0 \\ 0 & G_0 \end{bmatrix}.$$

By matrix inversion and multiplication we find the admittance parameters in terms of the scattering parameters as

$$Y_{11} = \frac{G_0}{\Sigma_{SY}}(1 + S_{22} - S_{11} - \Delta_S), \qquad (2.3.17a)$$

$$Y_{12} = -\frac{2}{\Sigma_{SY}} G_0 S_{12}, \qquad (2.3.17b)$$

$$Y_{21} = -\frac{2}{\Sigma_{SY}} G_0 S_{21}, \qquad (2.3.17c)$$

$$Y_{22} = \frac{G_0}{\Sigma_{SY}}(1 + S_{11} - S_{22} - \Delta_S) \qquad (2.3.17d)$$

where

$$\Sigma_{SY} = 1 + S_{11} + S_{22} + \Delta_S \qquad (2.3.18)$$

and

$$\Delta_S = S_{11}S_{22} - S_{12}S_{21}.$$

Conversion of the scattering matrix in a system of reference impedances to another system of reference impedances

The conversion of the scattering parameters from one system of reference impedances to another system of different reference impedances can be achieved in two steps. We first convert the scattering parameters to the impedance (or admittance) parameters, then the impedance (or admittance) parameters can be converted to the scattering parameters in a new system of reference impedances.

Example (2.3.1)

Using the conversion between the admittance parameters and the scattering parameters, find the scattering parameters of Example (2.1.2).

Solution

From the definition of the admittance parameters, the admittance parameters of a series impedance Z_a are given by $Y_{11} = Y_a$, $Y_{12} = -Y_a$, $Y_{21} = -Y_a$ and $Y_{22} = Y_a$. With $Z_a = 100 + j50$,

$$Y_a = 0.008 - j0.004 \ , \quad Y_{11} = Y_{22} = -Y_{12} = -Y_{22} = 0.008 - j0.040.$$

We now consider the two cases separately.

(I) $R_0 = 50$ or $G_0 = 0.02$ and hence $\Sigma_{YS} = 0.00072 - j0.00016$ and from (2.3.16)

$$S_{11} = S_{22} = 0.529 + j0.118,$$

$$S_{12} = S_{21} = 0.471 - j0.118.$$

(ii) $R_0 = 100$ or $G_0 = 0.01$ and hence $\Sigma_{YS} = 0.00026 - j0.00008$ and again from (2.3.16)

$$S_{11} = S_{22} = 0.351 + j0.108,$$

$$S_{12} = S_{21} = 0.649 - j0.108.$$

The above results are all identical to the results found in Example (2.1.2).

Example (2.3.2)

Using the conversion between the impedance parameters and the scattering matrix parameters, find the scattering matrix parameters of Example (2.1.4).

Solution

From the definition of impedance parameters the impedance parameters of a shunt impedance Z_b are given by

$$Z_{11} = Z_{12} = Z_{21} = Z_{22} = Z_b = 50 + j100.$$

(i) $R_0 = 50$, gives $\Sigma_{ZS} = 7,500 + j10,000$ and hence from (2.3.7)

$$S_{11} = S_{22} = -0.120 + j0.160,$$

$$S_{12} = S_{21} = 0.880 + j0.160.$$

(ii) $R_0 = 100$,   gives $\Sigma_{ZS} = 20,000 + j20,000$ and hence from (2.3.7)

$$S_{11} = S_{22} = -0.250 + j0.250,$$

$$S_{12} = S_{21} = 0.750 + j0.250.$$

The above results are all identical to the results found from Example (2.1.4).

Example (2.3.3)

From the definition of the impedance parameters, find the impedance parameters of a π-section as shown below

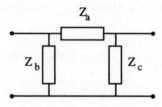

Solution

The following expressions can be written directly from the definition of the impedance parameters

$$Z_{11} = \frac{1}{Y_b + Y_d} = \frac{Z_b(Z_a + Z_c)}{Z}, \qquad Z_{12} = \frac{Y_e}{Y_c + Y_e} Z_b = \frac{Z_b Z_c}{Z},$$

$$Z_{21} = \frac{Y_d}{Y_b + Y_d} Z_c = \frac{Z_c Z_b}{Z}, \qquad Z_{22} = \frac{1}{Y_c + Y_e} = \frac{Z_c(Z_a + Z_b)}{Z}.$$

In the above equations $Y_d = 1/(Z_a + Z_c)$, $Y_e = 1/(Z_a + Z_b)$ and $Z = Z_a + Z_b + Z_c$. As expected $Z_{12} = Z_{21}$.

Example (2.3.4)

With $Z_a = 100 + j50$, $Z_b = 50 + j100$ and $Z_c = 100$ find the impedance parameters of Example (2.3.3).

Solution

Substituting into the expressions of the previous section, we have

$$Z = 100 + j50 + 50 + j100 + 100 = 250 + j150$$

and hence

$$Z_{11} = 54.4 + j57.35, \qquad Z_{12} = 32.35 + j20.6,$$

$$Z_{21} = 32.35 + j20.6, \qquad Z_{22} = 70.6 + j17.6.$$

Example(2.3.5)

Find the scattering parameters for the π-section of Example (2.3.4) for $R_0 = 50$.

Solution

Substituting for values of Z_{11}, Z_{12}, Z_{21} and Z_{22} into (2.3.7), and with $\Sigma_{ZS} = 10956 + j7426$, we find

$$S_{11} = 0.171 + j0.401, \qquad S_{12} = 0.290 - j0.008,$$

$$S_{21} = 0.290 - j0.008, \qquad S_{22} = 0.104 + j0.084.$$

Example (2.3.6)

From the definition of the impedance parameters find the impedance parameters of the two given L-sections, and using impedance to scattering parameters conversion, find the scattering parameters of the L-sections in a system of references R_0.

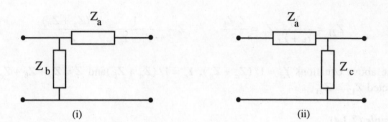

(i)   (ii)

Solution

(i) The impedance parameters can be written readily from the given definitions as $Z_{11} = Z_b$, $Z_{12} = Z_b$, $Z_{21} = Z_b$ and $Z_{22} = Z_a + Z_b$. Hence from (2.3.6),

$$\Sigma_{ZS} = (Z_b + R_0)(Z_a + Z_b + R_0) - Z_b^2 = (Z_b + R_0)(Z_a + R_0) + R_0 Z_b$$

and from (2.3.7)

$$S_{11} = [(Z_b - R_0)(Z_a + R_0) - R_0 Z_b] / \Sigma_{ZS},$$

$$S_{12} = 2R_0 Z_b / \Sigma_{ZS},$$

$$S_{21} = 2R_0 Z_b / \Sigma_{ZS},$$

$$S_{22} = [(Z_b + R_0)(Z_a - R_0) + R_0 Z_b] / \Sigma_{ZS}.$$

(ii) In this case, $Z_{11} = Z_a + Z_c$, $Z_{12} = Z_c$, $Z_{21} = Z_c$, $Z_{22} = Z_c$ and again

$$\Sigma_{ZS} = (Z_a + R_0)(Z_c + R_0) + R_0 Z_c$$

and finally from (2.3.7)

$$S_{11} = [(Z_c + R_0)(Z_a - R_0) + R_0 Z_c] / \Sigma_{ZS},$$

$$S_{12} = 2R_0 Z_c / \Sigma_{ZS},$$

$$S_{21} = 2R_0 Z_c / \Sigma_{ZS},$$

$$S_{22} = [(Z_c - R_0)(Z_a + R_0) - R_0 Z_c] / \Sigma_{ZS}.$$

Example (2.3.7)

Find the scattering parameters of the following L-sections when $R_0 = 50$. These are examples of some matching circuits that we shall discuss in the next chapter.

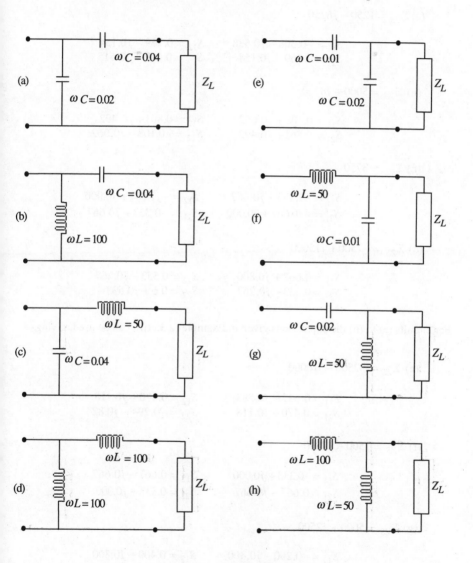

Examples of some matching circuits

Solution

For circuits of (a) to (d) we use the results of Example (2.3.6(i)) which gives

(a) $\Sigma_{ZS} = 1250 - j6250$

$$S_{11} = -0.308 - j0.538 \qquad S_{12} = 0.769 - j0.154$$
$$S_{21} = 0.769 - j0.154 \qquad S_{22} = 0.077 - j0.615$$

(b) $\Sigma_{ZS} = 5000 + j8750$

$$S_{11} = 0.108 + j0.062 \qquad S_{12} = 0.861 + j0.492$$
$$S_{21} = 0.861 + j0.492 \qquad S_{22} = -0.108 - j0.062$$

(c) $\Sigma_{ZS} = 3750$

$$S_{11} = -0.333 - j0.667 \qquad S_{12} = -j0.667 + j0.000$$
$$S_{21} = -j0.667 + j0.000 \qquad S_{22} = -0.333 + j0.667$$

(d) $\Sigma_{ZS} = -7500 + j15000$

$$S_{11} = 0.067 + j0.800 \qquad S_{12} = 0.533 - j0.267$$
$$S_{21} = 0.533 - j0.267 \qquad S_{22} = 0.6 + j0.533$$

For circuits (e) to (h) the expressions given in Example (2.3.6(ii)) can be used, giving

(e) $\Sigma_{ZS} = -2500 - j10000$

$$S_{11} = 0.647 - j0.588 \qquad S_{12} = 0.470 + j0.118$$
$$S_{21} = 0.470 + j0.118 \qquad S_{22} = -0.294 - j0.824$$

(f) $\Sigma_{ZS} = 7500 - j7500$

$$S_{11} = j0.333 + j0.000 \qquad S_{12} = 0.667 - j0.667$$
$$S_{21} = 0.667 - j0.667 \qquad S_{22} = 0.333 + j0.000$$

(g) $\Sigma_{ZS} = 5000 + j2500$

$$S_{11} = -0.200 - j0.400 \qquad S_{12} = 0.400 + j0.800$$
$$S_{21} = 0.400 + j0.800 \qquad S_{22} = 0.200 + j0.400$$

(h) $\Sigma_{ZS} = -2500 + j10000$

$$S_{11} = 0.647 + j0.588 \qquad S_{12} = 0.471 - j0.118$$
$$S_{21} = 0.471 - j0.118 \qquad S_{22} = -0.294 + j0.824$$

Example (2.3.8)

By finding the admittance and impedance parameters as an intermediate step, convert the scattering parameters of examples (2.1.2i) and (2.1.4i) from a system of reference impedances of 50 to 100, and hence verify the results of Example (2.1.2ii) and Example (2.1.4ii).

For Example (2.1.2i) with $R_0 = 50$, we have found

$$S_{11} = S_{22} = 0.529 + j0.118, \qquad S_{12} = S_{21} = 0.471 - j0.118.$$

This gives $\Delta_S = S_{11}S_{22} - S_{12}S_{21} = 0.058 + j0.236$ and

$$\Sigma_{SY} = 1 + S_{11} + S_{22} + \Delta_S = 2.116 + j0.472.$$

Hence from (2.3.17) we have

$$Y_{11} = Y_{22} = 0.008 - j0.004, \qquad Y_{12} = Y_{21} = -0.008 + j0.004.$$

With $R_0 = 100$, converting the admittance parameters to scattering parameters as in Example (2.3.1ii), we find

$$S_{11} = S_{22} = 0.351 + j0.108, \qquad S_{12} = S_{21} = 0.649 - j0.108$$

as in Example (2.1.2ii). For Example (2.1.4i) with $R_0 = 50$, we have found

$$S_{11} = S_{22} = -0.120 + j0.160, \qquad S_{12} = S_{21} = 0.880 + j0.160.$$

This gives $\Delta_S = S_{11}S_{22} - S_{12}S_{21} = -0.760 - j0.320$ and

$$\Sigma_{SZ} = 1 - S_{11} - S_{22} + \Delta_S = 0.480 - j0.640.$$

Hence from (2.3.8) we have, $Z_{11} = Z_{22} = Z_{12} = Z_{21} = 50 + j100.$

Converting the impedance   parameters to the scattering parameters as in Example (2.3.2ii) we find

$$S_{11} = S_{22} = -0.250 + j0.250, \qquad S_{12} = S_{21} = 0.750 + j0.250$$

as in Example (2.1.4ii).

2.4 The Wave Amplitudes of a General 2-port Network - Signal Flow Graphs

As we have discussed before, the incident and reflected amplitudes are not unique but are dependent on both the reference impedances and the impedance of the generators. We have given the definition of the incident and reflected wave amplitudes when the generator impedances were equal to the reference impedances. In this section we consider the modification of the wave amplitudes when the impedances of the generators are different from the given reference impedances.

In particular consider a 2-port network connecting a generator of emf E_S and internal impedance Z_S to a load of impedance Z_L, as shown in Fig. (2.4.1).

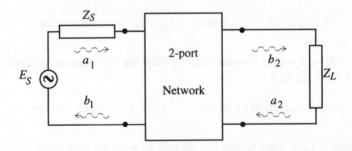

Fig. (2.4.1)

Assuming that the scattering parameters of the network are given in a system of reference impedances R_0, we require the wave amplitudes a_1, b_1, a_2 and b_2 of the given network, in the same system of reference impedances.

The wave amplitude a_1 is modified and can be written as

$$a_1 = b_S + \Gamma_S \, b_1 \tag{2.4.1}$$

where

$$b_S = \frac{\sqrt{R_0}E_S}{Z_S + R_0} = \frac{E_S}{2\sqrt{R_0}}(1 - \Gamma_S).$$

In (2.4.1) the first term on the right is the incident wave when $b_1 = 0$, or when port 1 is terminated by a resistance R_0. The second term represents the contribution to a_1 by b_1, partially reflected by the source impedance Z_S.

With no generator in the output port a_2 and b_2 are related as

$$a_2 = \Gamma_L b_2 \tag{2.4.2}$$

where Γ_L is the reflection coefficient of the load.

Finally from the definition of the scattering parameters the wave amplitudes are related by the expressions

$$b_1 = S_{11}a_1 + S_{12}a_2, \tag{2.4.3}$$

$$b_2 = S_{21}a_1 + S_{22}a_2. \tag{2.4.4}$$

The above four equations relate the four unknown wave amplitudes to the source parameter b_S. The solutions are given as

$$a_1 = b_S \frac{1 - \Gamma_L S_{22}}{(1 - \Gamma_S S_{11})(1 - \Gamma_L S_{22}) - \Gamma_S \Gamma_L S_{12}S_{21}}, \tag{2.4.5}$$

$$a_2 = b_S \frac{\Gamma_L S_{21}}{(1 - \Gamma_S S_{11})(1 - \Gamma_L S_{22}) - \Gamma_S \Gamma_L S_{12}S_{21}}, \tag{2.4.6}$$

$$b_1 = b_S \frac{S_{11} - \Gamma_L \Delta_S}{(1 - \Gamma_S S_{11})(1 - \Gamma_L S_{22}) - \Gamma_S \Gamma_L S_{12}S_{21}}, \tag{2.4.7}$$

$$b_2 = b_S \frac{S_{21}}{(1 - \Gamma_S S_{11})(1 - \Gamma_L S_{22}) - \Gamma_S \Gamma_L S_{12}S_{21}}. \tag{2.4.8}$$

The relations between the wave amplitudes may be represented by a 'Flow Graph' as shown in Fig. (2.4.2).

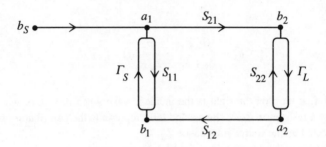

Fig. (2.4.2)

The above relations between the wave amplitudes a_1, a_2, b_1, b_2 and the wave amplitude b_S, can be readily derived from the Flow Graph using what is known as Mason's rule. In fact the rule can be applied to much more complicated situations, simplifying many algebraic manipulations.

Mason's rule

A Flow Graph consists of a number of nodes and directed branches, with nodes representing the wave amplitudes and branches the scattering parameters, similar to those given in Fig. (2.4.2). If we impress a wave of amplitude unity on any node, the wave amplitude at any other node can be found by the use of Mason's rule. To state this rule, however, we require the following definitions

(i) a 'path' is the product of all branches encountered moving in the indicated direction of each branch from any node to any other node.

(ii) a 'first order loop' is the product of all branches encountered, moving in the direction of the branch from any node and returning to the same node.

(iii) a 'second order loop' is the product of any two non-touching (not passing through a common node) first order loops.

(iv) a 'kth order loop' is the product of any k non-touching first order loops.

If a wave of unit amplitude is impressed at a particular node, the wave amplitude at any other node is given by the expression

$$\{P_1[1 - \Sigma L(1)^{(1)} + \Sigma L(2)^{(1)} - ....]$$

$$+P_2[1 - \Sigma L(1)^{(2)} + \Sigma L(2)^{(2)} - ....]$$

$$+....+P_K[1 - \Sigma L(1)^{(K)} + \Sigma L(2)^{(K)} - ....]+....\}$$

$$/[1 - \Sigma L(1) + \Sigma L(2) - ....] \qquad (2.4.9)$$

where P_1, P_2 .. P_K ... are the paths from the first to the second node. $\Sigma L(1)$, $\Sigma L(2)$, ... are the sums of the first, second and all higher order loops. $\Sigma L(1)^{(K)}$, $\Sigma L(2)^{(K)}$, ... denote the sums of the first order, second order and all higher order loops that do not touch the path P_K, from the first to the second node.

Application of the Mason's rule to 2-port network Flow Graphs

We now demonstrate the application of Mason's Rule by finding the values a_1, a_2, b_1 and b_2 and comparing them with those given in (2.4.5) to (2.4.8). In the Flow Graph (Fig. (2.4.2)), the first order loops are $\Gamma_S S_{11}$, $\Gamma_L S_{22}$ and $S_{21}\Gamma_L S_{12}\Gamma_S$. The only second order loop is $\Gamma_S S_{11}\Gamma_L S_{22}$ which is the product of the only non-touching loops $\Gamma_S S_{11}$ and $\Gamma_L S_{22}$. The denominator for all the expressions is therefore given by

$$1 - \Gamma_S S_{11} - \Gamma_L S_{22} - S_{21}\Gamma_L S_{12}\Gamma_S + \Gamma_S S_{11}\Gamma_L S_{22} =$$

$$(1 - \Gamma_S S_{11})(1 - \Gamma_L S_{22}) - \Gamma_S \Gamma_L S_{12}S_{21}$$

as in (2.4.5) to (2.4.8).

For the node a_1, we note that the only path from b_S to a_1 is through the branch of value 1, hence $P_1 = 1$. The only non-touching loop to this path is the first order loop $\Gamma_L S_{22}$. Hence the numerator of (2.4.9) is $b_S(1 - \Gamma_L S_{22})$ as in (2.4.5).

For the node a_2, there is again only one path from b_S of value $P_1 = 1 \times S_{21}\Gamma_L$. There is no non-touching loop to this path. The numerator of (2.4.9) is hence $b_S \Gamma_L S_{21}$ as in (2.4.6).

For the node b_1, there are two paths from b_S to b_1, with values of $P_1 = (1) \times S_{11}$ and $P_2 = (1) \times S_{21}\Gamma_L S_{12}$. The non-touching loop to the path P_1 is $\Gamma_L S_{22}$ and there is no non-touching loop to the path P_2. The numerator in (2.4.9) is therefore

$$b_S[S_{11}(1 - \Gamma_L S_{22}) + S_{21}\Gamma_L S_{12}] = b_S[S_{11} - \Gamma_L(S_{11}S_{22} - S_{21}S_{12})] = b_S[S_{11} - \Gamma_L \Delta_S]$$

as in (2.4.7).

For the node b_2, there is a path from b_S given by $P_1 = b_S S_{21}$ and there is no non-touching loop to this path. The numerator of (2.4.9) is $b_S S_{21}$ as in (2.4.8).

Example (2.4.1)

The scattering parameters of a 2-port network is measured in a system of reference impedances $\hat{Z}_1 = \hat{Z}_2 = 50$ and given as

$$S_{11} = 0.40 + j0.20, \qquad S_{12} = 0.10 + j0.05,$$

$$S_{21} = 3.00 - j0.50, \qquad S_{22} = 0.30 - j0.10.$$

The network is terminated at its input port by a generator of emf 5 and internal impedance $30 + j20$ and at its output port by a load of $70-j30$. Calculate a_1, a_2, b_1 and b_2.

Solution

$\Gamma_S = -0.176 + j0.294$ and $\Gamma_L = 0.216 - j0.196$. With $b_S = E_S(1-\Gamma_S)/(2\sqrt{R_0}) = 0.416 - j0.104$ and $(1-\Gamma_S S_{11})(1-\Gamma_L S_{22}) - \Gamma_S \Gamma_L S_{12} S_{21} = 1.09 - j\,0.022$, the wave amplitudes can be calculated from (2.4.5) to (2.4.8) as

$$a_1 = 0.374 - j0.053, \qquad b_1 = 0.191 + j0.029,$$

$$a_2 = 0.150 - j0.315, \qquad b_2 = 1.108 - j0.456.$$

Example (2.4.2)

Using Mason's rule, find the overall scattering parameters S_{11}^T, S_{12}^T, S_{21}^T, S_{22}^T of two cascaded networks with the scattering parameters S_{11}^1, S_{12}^1, S_{21}^1, S_{22}^1 and S_{11}^2, S_{12}^2, S_{21}^2, S_{22}^2.

Solution

The Flow Graph for the combined network can be drawn as shown. From the definition of the scattering parameters, S_{11}^T is the ratio of the wave amplitude b_1^1 to the wave amplitude a_1^1 and S_{21}^T is the ratio of the wave amplitude b_2^2 to that of the wave amplitude a_1^1. S_{22}^T and S_{12}^T can be similarly defined. To find the overall scattering parameters, therefore, we should identify the first and higher order loops, as well as the paths between the relevant nodes.

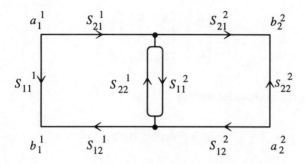

There is only one first order loop and no other higher order loops. The value of this loop is $S_{22}^1 S_{11}^2$.

There are two paths from a_1^1 to b_1^1: (i) S_{11}^1 with one non-touching loop $S_{22}^1 S_{11}^2$ and (ii) $S_{21}^1 S_{11}^2 S_{12}^1$ with no non-touching loops.

Similarly there are two paths from a_2^2 to b_2^2: (i) S_{22}^2 with non-touching loop $S_{22}^1 S_{11}^2$ and (ii)$S_{12}^2 S_{22}^1 S_{21}^2$ with no non-touching loops.

The paths from a_1^1 to b_2^2 and a_2^2 to b_1^1 have no non-touching loops and have values of $S_{21}^1 S_{21}^2$ and $S_{12}^2 S_{12}^1$ respectively. The scattering parameters of the combined network can hence be written as

$$S_{11}^T = \frac{S_{11}^1 - \Delta_S^1 S_{11}^2}{1 - S_{22}^1 S_{11}^2}, \qquad S_{12}^T = \frac{S_{12}^1 S_{12}^2}{1 - S_{22}^1 S_{11}^2},$$

$$S_{21}^T = \frac{S_{21}^1 S_{21}^2}{1 - S_{22}^1 S_{11}^2}, \qquad S_{22}^T = \frac{S_{22}^2 - \Delta_S^2 S_{22}^1}{1 - S_{22}^1 S_{11}^2}$$

where $\Delta_S^1 = S_{11}^1 S_{22}^1 - S_{12}^1 S_{21}^1$ and $\Delta_S^2 = S_{11}^2 S_{22}^2 - S_{12}^2 S_{21}^2$.

2.5 The Transfer Scattering Matrix

For cascaded 2-port networks it is convenient to define a new set of scattering parameters known as the Transfer Scattering Parameters or the T-parameters. The T-parameters are closely related to the scattering parameters (S-parameters), but relate the wave amplitudes a_1 and b_1 of the input port to the wave amplitudes a_2 and b_2 of the output port.

The scattering matrix **S** of a 2-port network was defined by the relation

$$\begin{bmatrix} b_1 \\ b_2 \end{bmatrix} = \begin{bmatrix} S_{11} & S_{12} \\ S_{21} & S_{22} \end{bmatrix} \begin{bmatrix} a_1 \\ a_2 \end{bmatrix}$$

which can be given in expanded form as

$$b_1 = S_{11}a_1 + S_{12}a_2, \tag{2.5.1a}$$

$$b_2 = S_{21}a_1 + S_{22}a_2. \tag{2.5.1b}$$

The T-parameters, however, are defined by the relation

$$\begin{bmatrix} a_1 \\ b_1 \end{bmatrix} = \begin{bmatrix} T_{11} & T_{12} \\ T_{21} & T_{22} \end{bmatrix} \begin{bmatrix} b_2 \\ a_2 \end{bmatrix}$$

or in the expanded form as

$$a_1 = T_{11}b_2 + T_{12}a_2, \tag{2.5.2a}$$

$$b_1 = T_{21}b_2 + T_{22}a_2. \tag{2.5.2b}$$

To find the T-parameters in terms of the S-parameters, Equation (2.5.1) is solved for a_1 and b_1 as

$$a_1 = -\frac{S_{22}}{S_{21}}a_2 + \frac{1}{S_{21}}b_2 \tag{2.5.3a}$$

and

$$b_1 = -\frac{\Delta_S}{S_{21}}a_2 + \frac{S_{11}}{S_{21}}b_2 \tag{2.5.3b}$$

where

$$\Delta_S = S_{11}S_{22} - S_{12}S_{21}.$$

Comparing (2.5.2) and (2.5.3), we identify the T-parameters as

$$T_{11} = \frac{1}{S_{21}}, \tag{2.5.4a}$$

$$T_{12} = -\frac{S_{22}}{S_{21}}, \tag{2.5.4b}$$

$$T_{21} = \frac{S_{11}}{S_{21}}, \tag{2.5.4c}$$

$$T_{22} = -\frac{\Delta_S}{S_{21}}. \tag{2.5.4d}$$

Similarly we can find the S-parameters in terms of T-parameters as

$$S_{11} = \frac{T_{21}}{T_{11}}, \tag{2.5.5a}$$

$$S_{12} = \frac{\Delta_T}{T_{11}}, \tag{2.5.5b}$$

$$S_{21} = \frac{1}{T_{11}}, \tag{2.5.5c}$$

$$S_{22} = -\frac{T_{12}}{T_{11}} \tag{2.5.5d}$$

where

$$\Delta_T = T_{11}T_{22} - T_{12}T_{21}.$$

To see the advantage of the T-parameters, consider a cascaded network of individual T-parameters T_{11}^1, T_{12}^1, T_{21}^1, T_{11}^2 and T_{11}^2, T_{12}^2, T_{21}^2, T_{22}^2 as shown in Fig. (2.5.1). For these two networks

$$\begin{bmatrix} a_1^1 \\ b_1^1 \end{bmatrix} = \begin{bmatrix} T_{11}^1 & T_{12}^1 \\ T_{21}^1 & T_{22}^1 \end{bmatrix} \begin{bmatrix} b_2^1 \\ a_2^1 \end{bmatrix}$$

and

$$\begin{bmatrix} a_1^2 \\ b_1^2 \end{bmatrix} = \begin{bmatrix} T_{11}^2 & T_{12}^2 \\ T_{21}^2 & T_{22}^2 \end{bmatrix} \begin{bmatrix} b_2^2 \\ a_2^2 \end{bmatrix}.$$

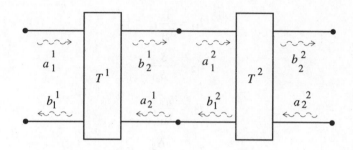

Fig. (2.5.1)

With $b_2^1 = a_1^2$ and $a_2^1 = b_1^2$ we have

$$\begin{bmatrix} a_1^1 \\ b_1^1 \end{bmatrix} = \begin{bmatrix} T_{11}^1 & T_{12}^1 \\ T_{21}^1 & T_{22}^1 \end{bmatrix} \begin{bmatrix} T_{11}^2 & T_{12}^2 \\ T_{21}^2 & T_{22}^2 \end{bmatrix} \begin{bmatrix} b_2^2 \\ a_2^2 \end{bmatrix}.$$

The above equation relates the input and output wave amplitudes of the combined network by a single transfer scattering matrix $\mathbf{T}^T$ given by the expression

$$\mathbf{T}^T = \mathbf{T}^1 \mathbf{T}^2.$$

Clearly no such direct relations can be found between the scattering parameters of the cascaded networks.

The above method can be extended to any number of cascaded 2-port networks. Hence the overall scattering matrix of a number of cascaded networks can be found by the three steps of:

(1) converting the scattering parameters of the individual networks to transfer parameters,

(2) derivation of the total transfer parameters by matrix multiplication,

(3) conversion of the transfer parameters to the usual scattering parameters.

Example (2.5.1)

Consider a π-section with series impedance $Z_a = 100 + j50$ and shunt impedances $Z_b = 50 + j100$ and $Z_c = 100$ as shown. Taking the π-section as three cascaded sections, find the T-parameters for each section and by matrix multiplication find the overall T-parameters. By converting the T-parameters to S-parameters find the scattering parameters of the given π-section. Assume the reference impedances R_0 to be 50.

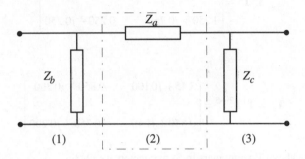

Solution

The scattering parameters of sections (1) and (2) have been already found in Examples (2.1.2) and (2.1.4). The scattering parameters of section (3) can be calculated similarly. We tabulate the results for each section, together with the T-parameters found using (2.5.4). Matrix multiplication of the T-matrices and the conversion of the T-parameters to the S-parameters give the required results.

	S-parameters	T-parameters
(1)	$S_{11} = -0.120 + j0.160$ $S_{12} = 0.880 + j0.160$ $S_{21} = 0.880 + j0.160$ $S_{22} = -0.120 + j0.160$	$T_{11} = 1.100 - j0.200$ $T_{12} = 0.100 - j0.200$ $T_{21} = -0.100 + j0.200$ $T_{22} = 0.900 + j0.200$
(2)	$S_{11} = 0.529 + j0.118$ $S_{12} = 0.471 - j0.118$ $S_{21} = 0.471 - j0.118$ $S_{22} = 0.529 + j0.118$	$T_{11} = 2.000 + j0.500$ $T_{12} = -1.000 - j0.500$ $T_{21} = 1.000 + j0.500$ $T_{22} = 0.000 - j0.500$
(3)	$S_{11} = -0.200$ $S_{12} = 0.800$ $S_{21} = 0.800$ $S_{22} = -0.200$	$T_{11} = 1.250$ $T_{12} = 0.250$ $T_{21} = -0.250$ $T_{22} = 0.750$

Hence

$$\mathbf{T^2 T^3} = \begin{bmatrix} 2.750 + j0.750 & -0.250 - j0.250 \\ 1.250 + j0.750 & 0.250 - j0.250 \end{bmatrix}$$

and

$$\mathbf{T}^T = \mathbf{T^1 T^2 T^3} = \begin{bmatrix} 3.45 + j0.100 & -0.350 - j0.300 \\ 0.550 + j1.40 & 0.350 - j0.200 \end{bmatrix}.$$

Converting the above T-parameters to S- parameters we obtain

$$S_{11} = 0.171 + j0.401, \quad S_{12} = 0.290 - j0.008,$$

$$S_{21} = 0.290 - j0.008, \quad S_{22} = 0.104 + j0.084$$

as found in Example (2.3.5)

Example (2.5.2)

Given two cascaded 2-port networks with the scattering parameters $(S_{11}^1, S_{12}^1, S_{21}^1, S_{22}^1)$ and $(S_{11}^2, S_{12}^2, S_{21}^2, S_{22}^2)$, find the overall scattering parameters of the combined network.

Solution

Denoting the elements of the matrix $\mathbf{T}^1$ of section 1 by $T_{11}^1, T_{12}^1, T_{21}^1$ and T_{22}^1 and the elements of the matrix $\mathbf{T}^2$ of section 2 by $T_{11}^2, T_{12}^2, T_{21}^2$ and T_{22}^2, we can write the scattering matrix of the cascaded network $\mathbf{T}^T$ as

$$\mathbf{T}^T = \mathbf{T^1 T^2} = \begin{bmatrix} T_{11}^1 T_{11}^2 + T_{12}^1 T_{21}^2 & T_{11}^1 T_{12}^2 + T_{12}^1 T_{22}^2 \\ T_{21}^1 T_{11}^2 + T_{22}^1 T_{21}^2 & T_{21}^1 T_{12}^2 + T_{22}^1 T_{22}^2 \end{bmatrix}.$$

The overall scattering parameters using (2.5.5) are easily found as

$$S_{11}^T = \frac{T_{21}^1 T_{11}^2 + T_{22}^1 T_{21}^2}{T_{11}^1 T_{11}^2 + T_{12}^1 T_{21}^2}, \qquad S_{12}^T = \frac{\Delta_T^1 \Delta_T^2}{T_{11}^1 T_{11}^2 + T_{12}^1 T_{21}^2},$$

$$S_{21}^T = \frac{1}{T_{11}^1 T_{11}^2 + T_{12}^1 T_{21}^2}, \qquad S_{22}^T = -\frac{T_{11}^1 T_{12}^2 + T_{12}^1 T_{22}^2}{T_{11}^1 T_{11}^2 + T_{12}^1 T_{21}^2}.$$

Substituting for T-parameters in terms of the S-parameters, using (2.5.4) we have finally

$$S_{11}^T = \frac{S_{11}^1 - \Delta_S^1 S_{11}^2}{1 - S_{22}^1 S_{11}^2}, \qquad S_{12}^T = \frac{S_{12}^1 S_{12}^2}{1 - S_{22}^1 S_{11}^2},$$

$$S_{21}^T = \frac{S_{21}^1 S_{21}^2}{1 - S_{22}^1 S_{11}^2}, \qquad S_{22}^T = \frac{S_{22}^2 - \Delta_S^2 S_{22}^1}{1 - S_{22}^1 S_{11}^2}$$

where $\Delta_S^1 = S_{11}^1 S_{22}^1 - S_{12}^1 S_{21}^1$ and $\Delta_S^2 = S_{11}^2 S_{22}^2 - S_{12}^2 S_{21}^2$ as found in Example (2.4.2).

Example (2.5.3)

The scattering parameters of a 2-port network in a system of reference impedances $Z_1 = Z_2 = R_0$ are given by S_{11}, S_{12}, S_{21} and S_{22}. Find the scattering parameters of the 2-port network when it is cascaded by two sections of transmission line of characteristic impedance $Z = R_0$ and lengths l_1 and l_2 as shown.

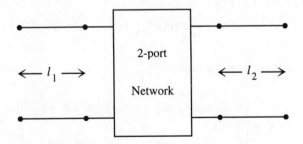

Solution

Assuming that the phase delay per unit length for both transmission lines to be β and with $\theta_1 = \beta l_1$ and $\theta_2 = \beta l_2$, the scattering matrices for the two transmission line sections are given as

$$\mathbf{S}^1 = \begin{bmatrix} 0 & \exp(-j\theta_1) \\ \exp(-j\theta_1) & 0 \end{bmatrix}, \qquad \mathbf{S}^2 = \begin{bmatrix} 0 & \exp(-j\theta_2) \\ \exp(-j\theta_2) & 0 \end{bmatrix}.$$

From scattering matrices, we can write T-matrices as

$$\mathbf{T}^1 = \begin{bmatrix} \exp(j\theta_1) & 0 \\ 0 & \exp(-j\theta_1) \end{bmatrix}, \qquad \mathbf{T}^2 = \begin{bmatrix} \exp(j\theta_2) & 0 \\ 0 & \exp(-j\theta_2) \end{bmatrix}$$

and

$$\mathbf{T}^{Net} = \begin{bmatrix} \dfrac{1}{S_{21}} & -\dfrac{S_{22}}{S_{21}} \\ \dfrac{S_{11}}{S_{21}} & -\dfrac{\Delta_S}{S_{21}} \end{bmatrix}$$

where $\mathbf{T}^{Net}$ refers to the transfer scattering matrix of the network.

For the cascaded 2-port network,

$$\mathbf{T}^T = \mathbf{T}^1 \mathbf{T}^{Net} \mathbf{T}^2 = \begin{bmatrix} \dfrac{1}{S_{21}}\exp[j(\theta_1+\theta_2)] & -\dfrac{S_{22}}{S_{21}}\exp[j(\theta_1-\theta_2)] \\ \dfrac{S_{11}}{S_{21}}\exp[j(\theta_2-\theta_1)] & -\dfrac{\Delta_S}{S_{21}}\exp[-j(\theta_1+\theta_2)] \end{bmatrix}$$

or by conversion to scattering parameters

$$\mathbf{S}^T = \begin{bmatrix} S_{11}\exp(-j2\theta_1) & S_{12}\exp[-j(\theta_1+\theta_2)] \\ S_{21}\exp[-j(\theta_1+\theta_2)] & S_{22}\exp(-j2\theta_2) \end{bmatrix}.$$

Chapter 3

Impedance Transforming Properties of a 2-port Network

A 2-port network acts as an impedance transformer. If a 2-port network is connected to a load at port 2, the impedance seen at port 1 is, in general, different from the load impedance. In terms of the reflection coefficients, the load reflection coefficient Γ_L is transformed by the network into the input reflection coefficient Γ_{IN}. Similarly the source reflection coefficient Γ_S is transformed into the output reflection coefficient Γ_{OT}. In this chapter the impedance transforming properties of networks are discussed in terms of the network scattering parameters. The matching networks that produce specific impedance transformations are also briefly examined. The complex plane representation of the impedances and reflection coefficients facilitates these considerations and is first discussed.

3.1 Complex Plane Representation of the Reflection Coefficients

The source and load reflection coefficients Γ_S and Γ_L, and the input and output reflection coefficients Γ_{IN} and Γ_{OT}, are in general complex and can be represented as points in a complex plane.

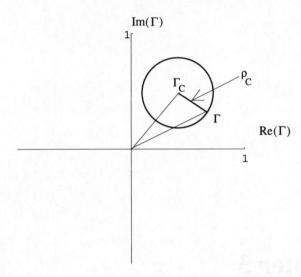

Fig. (3.1.1) Γ-plane

Any circle in a general Γ-plane, as shown in Fig. (3.1.1), has an equation in the form of

$$\left|\Gamma - \Gamma_C\right|^2 = \rho\,_C^2 \tag{3.1.1}$$

or

$$\Gamma\Gamma^* - (\Gamma_C\Gamma^* + \Gamma_C^*\Gamma) = \rho\,_C^2 - \left|\Gamma_C\right|^2 \tag{3.1.2}$$

where Γ_C represents the position of the center and ρ_C is the radius of the circle. Conversely, any expression in the form of (3.1.2) represents a circle in the complex Γ-plane.

As an example, consider the real and imaginary parts of a complex parameter z representing a normalized complex impedance, given by the expression

$$z = r + jx = \frac{1+\Gamma}{1-\Gamma} \tag{3.1.3}$$

From the above relation, the real part of z, denoted by r, is given as

$$r = \frac{1}{2}(z+z^*) = \frac{1}{2}\left[\frac{1+\Gamma}{1-\Gamma} + \frac{1+\Gamma^*}{1-\Gamma^*}\right]$$

or after rearrangement

$$\Gamma\Gamma^* - \left[\frac{r}{1+r}\Gamma^* + \frac{r}{1+r}\Gamma\right] = \frac{1-r}{1+r}. \qquad (3.1.4)$$

Comparing (3.1.2) and (3.1.4), we conclude that the locus of the points in the Γ-plane that represents a constant value of $r = \mathrm{Re}(z)$, is a circle with center on the real axis given by

$$\Gamma_{Cr} = \frac{r}{1+r} \qquad (3.1.5)$$

and radius

$$\rho_{Cr} = \left|\frac{1}{1+r}\right|. \qquad (3.1.6)$$

The circles for values of $r = 0$, ± 0.5, ± 1 and ± 2 are shown in Fig. (3.1.2b). Here the circle with heavy line is the circle $|\Gamma| = 1$.

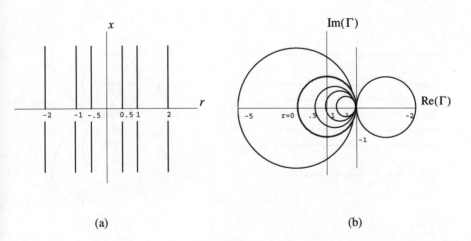

(a)   (b)

Fig. (3.1.2)

Similarly the imaginary value of z can be written as

$$x = \frac{1}{2j}(z - z^*) = \frac{1}{2j}\left[\frac{1+\Gamma}{1-\Gamma} - \frac{1+\Gamma^*}{1-\Gamma^*}\right]$$

or after rearrangement

$$\Gamma\Gamma^* - \left[\frac{x+j}{x}\Gamma^* + \frac{x-j}{x}\Gamma\right] = -1. \tag{3.1.7}$$

Comparing (3.1.2) and (3.1.7), we conclude that the loci of the points on the Γ-plane representing constant values of $x = \text{Im}(z)$, are circles with centers on the line

$$\Gamma_{Cx} = 1 + \frac{j}{x} \tag{3.1.8}$$

and radii

$$\rho_{Cx} = \left|\frac{1}{x}\right|. \tag{3.1.9}$$

Circles for $x = \pm0.5, \pm1$ and ±2 are shown in Fig. (3.1.3b). Here again the circle with heavy line is the circle $|\Gamma| = 1$.

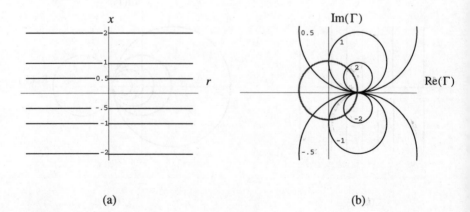

(a)   (b)

Fig. (3.1.3)

We can also consider the complex parameter y given by

$$y = g + jb = \frac{1}{z} = \frac{1-\Gamma}{1+\Gamma}. \tag{3.1.10}$$

Here the loci of the points of constant g and b are circles in the Γ-plane, and provided that we reverse the sense of both real and imaginary axes, they are identical to those for r and x.

As can be seen from Fig. (3.1.2b) the circles of constant r with r negative lie outside the $|\Gamma| = 1$ circle. Such circles in practice may extend beyond the limits of the page representing the complex plane.

For negative values of r it is more convenient, therefore, to plot the circles of constant r or x on a Δ-plane with Δ defined by the expression

$$\Delta = \frac{1}{\Gamma^*} \tag{3.1.11}$$

The normalized impedance z in terms of Δ is given as

$$z = -\frac{1+\Delta^*}{1-\Delta^*} \tag{3.1.12}$$

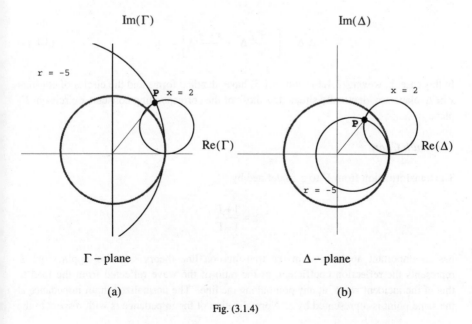

Γ – plane

(a)

Δ – plane

(b)

Fig. (3.1.4)

Again we can write

$$r = \frac{1}{2}(z + z^*) = -\frac{1}{2}\left[\frac{1+\Delta^*}{1-\Delta^*} + \frac{1+\Delta}{1-\Delta}\right]$$

or after rearrangement

$$\Delta \Delta^* - \left[\frac{-r}{1-r}\Delta^* + \frac{-r}{1-r}\Delta\right] = \frac{1+r}{1-r}. \tag{3.1.13}$$

Changing r to $-r$ in (3.1.13), this equation assumes the exact form of (3.1.4). It is clear, therefore, that the circles of constant r, with r positive are now outside the circle of unit radius in Δ-plane, while with r negative, the circles of constant r are inside the circle of unit radius. A constant r circle with r negative is demonstrated in a Γ-plane and a Δ-plane in Fig. (3.1.4a) and Fig. (3.1.4b) respectively.

Similarly the imaginary part of z can be written as

$$x = \frac{1}{2j}(z - z^*) = -\frac{1}{2j}\left[\frac{1+\Delta^*}{1-\Delta^*} - \frac{1+\Delta}{1-\Delta}\right]$$

or

$$\Delta \Delta^* - \left[\frac{x+j}{x}\Delta^* + \frac{x-j}{x}\Delta\right] = -1. \tag{3.1.14}$$

In this case, however, (3.1.14) and (3.1.7) have identical forms and the circles of constant x in Δ-plane are identically situated as those of the corresponding constant x circles in Γ-plane.

The Smith Chart

The transformation from Γ to z as defined by

$$z = \frac{1+\Gamma}{1-\Gamma}$$

has an important application in the transmission line theory. In this application, Γ represents the reflection coefficient, or the ratio of the wave reflected from the load to that of the incident wave, at any point along the line. The normalized input impedance at the same point is represented by z. Normalization of the impedance is with respect to the characteristic impedance of the line. The circles of constant r and x are the circles of constant normalized resistance and reactance (real and imaginary part of z) respectively.

With numerous practical applications of such mappings of the impedance-plane into the reflection coefficient-plane, the plots are given a particular name, the 'Smith Chart', after the person who originally introduced the Chart.

As we move along the transmission line, there is a change in the phase of the reflection coefficient, while its magnitude remains constant (transmission line assumed lossless). The fraction of the wavelength moved towards the generator or the load from a reference plane is usually indicated on the Chart. Hence the magnitude and phase of the reflection coefficient at any point on the line can easily be found, and the corresponding input resistance and reactance determined. For a lossless line, any two points, half a wavelength apart, have equal reflection coefficients and are hence represented by a single point on the Chart.

The Smith Chart is usually given for the values of the reflection coefficients with magnitude $|\Gamma| \leq 1$, as shown in Fig (3.1.5). However, according to the above discussion, the same region of the Chart can also be used for values of $|\Gamma| > 1$, provided that the complex plane is considered as a Δ-plane ($\Delta = 1/\Gamma^*$) rather than a Γ-plane.

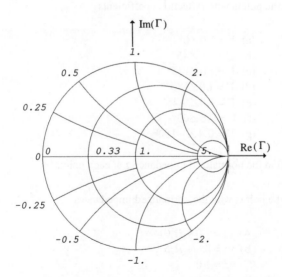

Fig. (3.1.5) Impedance or Admittance Chart

As we have seen the circles of constant normalized conductance g and susceptance b coincide with the circles of the same constant normalized resistance r and reactance x, if the sense of both real and imaginary axes in the Γ-plane are reversed. Hence a normalized impedance chart can be used as an admittance chart, provided that each point on the chart, representing a reflection coefficient Γ, is reflected about the origin to a new point with corresponding reflection coefficient of $-\Gamma$.

The use of the Smith Chart is not confined to that of transmission lines. The Chart can be used whenever a transformation between two complex planes such as z and Γ-planes in the form of

$$z = \frac{1+\Gamma}{1-\Gamma} \quad \text{and} \quad \Gamma = \frac{z-1}{z+1}$$

is required. For 2-port networks each pair of (Z_S and Γ_S), (Z_L and Γ_L), (Z_{IN} and Γ_{IN}) and (Z_{OT} and Γ_{OT}) are similarly related and hence the Smith Chart can be used for the calculation of the impedances from the respective reflection coefficients (or the reverse).

Example (3.1.1)

On the Smith Chart of Fig. (3.1.5)

(a) Locate the points with reflection coefficients

(a) $\Gamma = 1.0\angle 0°$
(b) $\Gamma = 1.0\angle 45°$
(c) $\Gamma = 1.0\angle 135°$
(d) $\Gamma = 1.0\angle 180°$
(e) $\Gamma = 1.0\angle -45°$
(f) $\Gamma = 0.6\angle 60°$
(g) $\Gamma = 0.4\angle -150°$

and hence find the normalized impedance z at each point.

(b) Locate the points with the normalized impedances

(A) $z = 1.0 + j1.0$
(B) $z = 0.0 + j1.5$
(C) $z = 1.0$
(D) $z = 0.0$
(E) $z = \infty$
(F) $z = 0.5 - j0.6$
(G) $z = 0.8$

Hence find the reflection coefficients Γ for each case.

Solution

The above points are shown on the Smith Chart plots and the impedances and reflection coefficients are given below.

(a)	$z = \infty$	(A)	$\Gamma = 0.447 \angle 63.4°$
(b)	$z = j2.41$	(B)	$\Gamma = 1.0 \angle 67.5°$
(c)	$z = j0.414$	(C)	$\Gamma = 0.0$
(d)	$z = 0.0$	(D)	$\Gamma = 1.0 \angle 180°$
(e)	$z = -j2.41$	(E)	$\Gamma = 1.0$
(f)	$z = 0.842 + j1.37$	(F)	$\Gamma = 0.483 \angle -108°$
(g)	$z = 0.45 - j0.216$	(G)	$\Gamma = 0.111 \angle 180°$

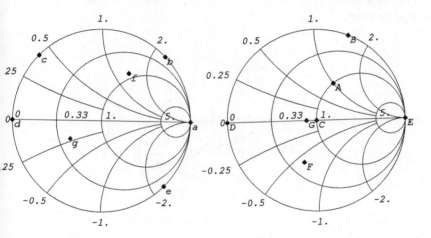

Example (3.1.2)

Two impedances $Z_L = 30 + j60$ and $Z = 120 - j30$ are connected by a transmission line of characteristic impedance $Z_0 = 50$ and length $\lambda/8$ as shown. Find the input impedance Z_{IN} of the network.

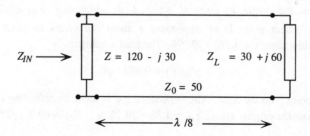

Solution

Normalizing the impedances with respect to $Z_0 = 50$, we have

$$z_L = 0.60 + j1.20$$

and

$$z = 2.40 - j0.60$$

Locating z_L on the Smith Chart we find point A with reflection coefficient $\Gamma_L = 0.632 \angle 71.6°$. Moving towards the generator by a distance $\lambda/8$, the magnitude of the reflection coefficient remains constant but its phase decreases by $2 \times 2\pi(\lambda/8)/\lambda = \pi/2 = 90°$. Hence $\Gamma_{IN} = 0.632 \angle -18.4°$ and $z'_{IN} = 3.0 - j2.0$ at point B on the Chart.

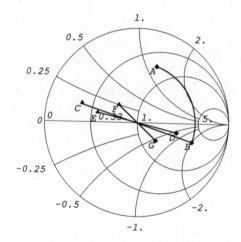

The normalized impedance z is in parallel with the normalized impedance z'_{IN} and hence we have to convert both normalized impedances to normalized admittances. Reflecting point B about the origin to point C we find an admittance $y'_{IN} = 0.231 + j0.154$. Similarly we reflect point D of impedance z about the origin to point E giving a normalized admittance $y = 0.392 + j0.098$. The total admittance is

$$y_{IN} = y'_{IN} + y = 0.623 + j0.252$$

which is the point F on the chart. The impedance z_{IN} is found by reflecting back point F to point G about the origin. Hence $z_{IN} = 1.38 - j0.56$ or $Z_{IN} = 69.0 - j27.9$.

3.2 Input and Output Impedances and Reflection Coefficients

The input reflection coefficient of a 2-port network (Fig. (3.2.1)) for a particular load impedance Z_L, is defined as the ratio of the reflected wave to the incident wave for the input port. The reflection coefficient is independent of the source impedance Z_S.

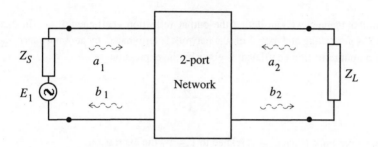

Fig. (3.2.1)

The reflection coefficient can be easily found from (2.4.5) and (2.4.7) as

$$\Gamma_{IN} = \frac{S_{11} - \Delta_S \Gamma_L}{1 - S_{22}\Gamma_L}.$$ (3.2.1)

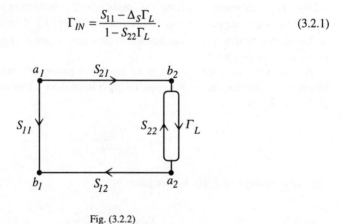

Fig. (3.2.2)

Alternatively we can consider the corresponding Flow Graph of Fig. (3.2.2). In this Flow Graph we have assumed $Z_S = R_0$ for convenience (since the reflection coefficient is independent of the source impedance). From a_1 to b_1 we have both a direct path and a path through b_2 and a_2. The direct path from a_1 to b_1 has a value S_{11} and a non-touching loop $S_{22} \Gamma_L$. The second path has a value $S_{21} \Gamma_L S_{12}$, and no non-touching loops. Using Mason's rule we can readily verify that the ratio b_1 / a_1 is as given by (3.2.1).

The input impedance Z_{IN} is related to Γ_{IN} by the expression

$$Z_{IN} = \frac{1 + \Gamma_{IN}}{1 - \Gamma_{IN}} R_0.$$

In a similar manner we can define the output reflection coefficient Γ_{OT}. In this case source 2 is generating and port 1 of the network is terminated by an impedance Z_S.

The output reflection coefficient is given by the expression

$$\Gamma_{OT} = \frac{S_{22} - \Delta_S \Gamma_S}{1 - S_{11} \Gamma_S} \tag{3.2.2}$$

and the output impedance Z_{OT} is related to Γ_{OT} by the expression

$$Z_{OT} = \frac{1 + \Gamma_{OT}}{1 - \Gamma_{OT}} R_0.$$

As can be seen from (3.2.1), a 2-port network acts as an impedance (or reflection coefficient) transformer, changing the impedance Z_L (reflection coefficient Γ_L) of the load to the input impedance Z_{IN} (reflection coefficient Γ_{IN}). Networks of appropriate scattering parameters can be synthesized to give any required impedance transformation over a range of frequencies.

Alternatively, for a network of given scattering parameters, we may wish to find an appropriate load impedance for a required input impedance. This can be found by solving (3.2.1) for Γ_L as

$$\Gamma_L = \frac{S_{11} - \Gamma_{IN}}{\Delta_S - S_{22} \Gamma_{IN}}. \tag{3.2.3}$$

Similarly solving (3.2.2) for Γ_S, we have

$$\Gamma_S = \frac{S_{22} - \Gamma_{OT}}{\Delta_S - S_{11} \Gamma_{OT}}. \tag{3.2.4}$$

For an input impedance equal to R_0, $\Gamma_{IN} = 0$ and assuming $S_{12} \neq 0$,

$$\Gamma_L = S_{11} / \Delta_S.$$

Similarly for an output impedance equal to R_0, $\Gamma_{OT} = 0$ and again assuming $S_{12} \neq 0$

$$\Gamma_S = S_{22} / \Delta_S.$$

Example (3.2.1)

A 2-port network is terminated by an impedance $Z_L = 50 + j25$. If the scattering parameters of the network are given as

$$S_{11} = 0.60\angle 30°, \qquad S_{12} = 0.02\angle 15°,$$

$$S_{21} = 2.00\angle 0°, \qquad S_{22} = 0.80\angle -45°$$

find the input impedance of the circuit.

Solution

The input reflection coefficient was given by (3.2.1) as

$$\Gamma_{IN} = (S_{11} - \Gamma_L \Delta_S) / (1 - \Gamma_L S_{22})$$

with $\Delta_S = S_{11}S_{22} - S_{12}S_{21} = 0.425 - j0.135$ and $1 - \Gamma_L S_{22} = 0.833 - j0.100$, we have

$$\Gamma_{IN} = 0.518 + j0.311 \text{ or } Z_{IN} = 96.4 + j94.6.$$

Example (3.2.2)

For the network of the Example (3.2.1), if the source impedance is $Z_S = 30 - j45$, calculate the output impedance of the circuit.

Solution

The output impedance is given by (3.2.2) as

$$\Gamma_{OT} = (S_{22} - \Gamma_S \Delta_S) / (1 - \Gamma_S S_{11})$$

with $\Delta_S = S_{11}S_{22} - S_{12}S_{21} = 0.425 - j0.135$ and $1 - \Gamma_S S_{11} = 0.814 + j0.262$, we have

$$\Gamma_{OT} = 0.567 - j0.591 \text{ or } Z_{OT} = 30.7 - j110.1.$$

Example (3.2.3)

For Example (3.1.2) find the scattering parameters of the combined network and hence find the input impedance and compare with the result already found.

Solution

The scattering parameters of the shunt impedance $Z_L = 30 + j60$ can be found as in Example (2.1.3) and are given as

$$S_{11} = S_{22} = -0.208 + j0.226,$$

$$S_{12} = S_{21} = 0.792 + j0.226.$$

Converting to transfer scattering parameters

$$T_{11} = 1.167 - j0.333, \qquad T_{12} = 0.167 - j0333,$$

$$T_{21} = -0.167 + j0.333, \qquad T_{22} = 0.833 + j0.333.$$

The scattering parameters of the transmission line can be found by considering that a matched transmission line of length ℓ introduces a phase delay of $2\pi\ell/\lambda$ for the waves traveling in either direction. There are, however, no wave reflections at either end of the line. Hence for $\ell = \lambda/8$ we can easily verify that

$$S_{11} = S_{22} = 0,$$

$$S_{12} = S_{21} = e^{-j\pi/4} = \sqrt{2}(1-j)/2.$$

Converting to transfer scattering parameters

$$T_{11} = 0.707 + j0.707, \qquad T_{12} = 0,$$

$$T_{21} = 0, \qquad\qquad T_{22} = 0.707 - j0.707.$$

The scattering parameters of the shunt impedance $Z = 120 - j30$ are

$$S_{11} = S_{22} = -0.165 - j0.034,$$

$$S_{12} = S_{21} = 0.835 - j0.034.$$

Converting to the transfer scattering parameters

$$T_{11} = 1.196 + j0.049, \qquad T_{12} = 0.196 + j0.049,$$

$$T_{21} = -0.196 - j0.049, \qquad T_{22} = 0.804 - j0.049.$$

Matrix multiplication of the three transfer scattering matrices gives

$$T_{11}^T = 1.246 + j0.832, \qquad T_{12}^T = 0.608 - j0.153,$$

$$T_{21}^T = -0.067 + j0.111, \qquad T_{22}^T = 0.571 - j0.319.$$

and hence

$$S_{11}^T = 0.004 + j0.086, \qquad S_{12}^T = 0.555 - j0.371,$$

$$S_{21}^T = 0.555 - j0.371, \qquad S_{22}^T = -0.281 + j0.310.$$

The network is open-circuited at the output ($\Gamma_L = 1$), thus its input reflection coefficient is given by

$$\Gamma_{IN} = S_{11}{}^T + \frac{S_{12}{}^T S_{21}{}^T}{1 - S_{22}{}^T}$$

or $\Gamma_{IN} = 0.276 \angle -43°$ and hence $Z_{IN} = 69 - j28$ as found before.

3.3 (Γ_{IN}, Γ_L) and (Γ_{OT}, Γ_S) Mapping

In the previous section we have related the input reflection coefficient of a 2-port network to the reflection coefficient of the load termination. The relation between the output reflection coefficient and the source reflection coefficient was also given. In this section we further investigate these relations, using a similar technique of transformation or mapping between two complex planes as given in section (3.1). Here the mapping is between the points in the input and load complex reflection coefficient planes and the output and source complex reflection coefficient planes.

For a 2-port network (Fig. (3.3.1)), as given by (3.2.1) to (3.2.4), the following relations between different reflection coefficients hold:

$$\text{(a)} \quad \Gamma_{IN} = \frac{S_{11} - \Delta_S \Gamma_L}{1 - S_{22} \Gamma_L} \qquad\qquad \text{(b)} \quad \Gamma_L = \frac{S_{11} - \Gamma_{IN}}{\Delta_S - S_{22} \Gamma_{IN}} \qquad (3.3.1)$$

$$\text{(a)} \quad \Gamma_{OT} = \frac{S_{22} - \Delta_S \Gamma_S}{1 - S_{11} \Gamma_S} \qquad\qquad \text{(b)} \quad \Gamma_S = \frac{S_{22} - \Gamma_{OT}}{\Delta_S - S_{11} \Gamma_{OT}} \qquad (3.3.2)$$

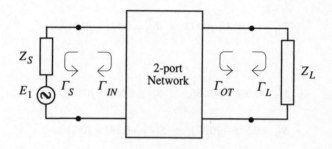

Fig. (3.3.1)

Equations (3.3.1a) and (3.3.1b) can be considered as representing a transformation or mapping between the points in the Γ_L-plane and Γ_{IN}-plane. Similarly Equations (3.3.2a) and (3.3.2b) represent a mapping between the Γ_{OT} and Γ_S planes.

Both (3.3.1) and (3.3.2) have the general form of

$$\bar{\gamma} = \frac{a + b\gamma}{c + d\gamma} \qquad (3.3.3)$$

where a, b, c and d are complex numbers.

Writing $\bar{\gamma}$ as

$$\bar{\gamma} = \bar{\rho}\, e^{j\bar{\tau}}$$

we can map all circles of constant $\bar{\rho}$ and lines of constant $\bar{\tau}$ in the $\bar{\gamma}$-plane to a set of corresponding circles in the γ-plane. To verify the above statement we write

$$\bar{\gamma}\bar{\gamma}^* = \bar{\rho}^2 = \frac{a + b\gamma}{c + d\gamma} \cdot \frac{a^* + b^*\gamma^*}{c^* + d^*\gamma^*}$$

which after expansion and rearrangement reduces to

$$\gamma\gamma^* - (\gamma_{C\overline{\rho}}^*\gamma + \gamma_{C\overline{\rho}}\gamma^*) = \rho_{C\overline{\rho}}^2 - |\gamma_{C\overline{\rho}}|^2 \tag{3.3.4}$$

where

$$\gamma_{C\overline{\rho}} = \frac{ab^* - \overline{\rho}^2 d^* c}{\overline{\rho}^2 |d|^2 - |b|^2} \tag{3.3.5}$$

and

$$\rho_{C\overline{\rho}} = \frac{\overline{\rho} \,|ad - bc|}{|\overline{\rho}^2 |d|^2 - |b|^2|} . \tag{3.3.6}$$

Comparing (3.3.4) with (3.1.2), it is clear that (3.3.4) represents a family of circles in the γ-plane with centers at $\gamma_{C\overline{\rho}}$ and radius $\rho_{C\overline{\rho}}$.

At the center of the $\overline{\gamma}$-plane $\overline{\rho} = 0$ and the corresponding circle in the γ-plane reduces to a point at $\gamma = -a/b$. Similarly when $\overline{\rho} \to \infty, \rho_{C\overline{\rho}} \to 0$ and hence the points at infinity in the $\overline{\gamma}$-plane map to a point at $-c/d$ in the γ-plane.

As we shall see in the next chapter, a circle of particular interest is the circle of $\overline{\rho} \to 1$ with the center and radius given as

$$\gamma_{C\overline{\rho}} \Big|_{\overline{\rho} \to 1} = \frac{ab^* - d^* c}{|d|^2 - |b|^2}, \tag{3.3.7}$$

$$\rho_{C\overline{\rho}} \Big|_{\overline{\rho} \to 1} = \frac{|ad - bc|}{\left||d|^2 - |b|^2\right|} . \tag{3.3.8}$$

Considering the lines of constant $\overline{\tau}$, let

$$\overline{T} = \tan \overline{\tau}$$

then

$$\overline{T} = \frac{\overline{\gamma} - \overline{\gamma}^*}{j(\overline{\gamma} + \overline{\gamma}^*)} = -j \frac{[(a + b\gamma)/(c + d\gamma)] - [(a^* + b^*\gamma^*)/(c^* + d^*\gamma^*)]}{[(a + b\gamma)/(c + d\gamma)] + [(a^* + b^*\gamma^*)/(c^* + d^*\gamma^*)]} .$$

Again after expansion and rearrangement we have

$$\gamma\gamma^* - (\gamma_{C\overline{\tau}}^*\gamma + \gamma_{C\overline{\tau}}\gamma^*) = \rho_{C\overline{\tau}}^2 - |\gamma_{C\overline{\tau}}|^2 \tag{3.3.9}$$

where

$$\gamma_{C\bar{\tau}} = -\frac{(b^*c + ad^*)\overline{T} - j(b^*c - ad^*)}{\left|(bd^* + b^*d)\overline{T} + j(bd^* - b^*d)\right|} \tag{3.3.10}$$

and

$$\rho_{C\bar{\tau}} = \frac{\left|bc - ad\right|\sqrt{(\overline{T}^2 + 1)}}{\left|(bd^* + b^*d)\overline{T} + j(bd^* - b^*d)\right|} \tag{3.3.11}$$

Hence all lines of constant $\bar{\tau}$ in $\bar{\gamma}$-plane map into circles of radius $\rho_{C\bar{\tau}}$ and centers $\gamma_{C\bar{\tau}}$ in the γ-plane.

As all the radial lines of constant $\bar{\tau}$ pass through the center of the $\bar{\gamma}$-plane, they all pass through the point $\gamma = -a/b$ in the γ-plane. These radial lines also extend to infinity in the $\bar{\gamma}$-plane and hence converge to the point $\gamma = -c/d$ in the γ-plane.

A typical mapping of the $\bar{\gamma}$-plane to the γ-plane is shown in Fig. (3.3.2).

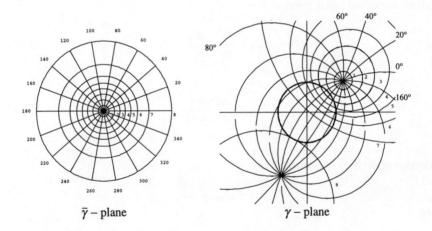

$$\bar{\gamma} - \text{plane} \qquad\qquad \gamma - \text{plane}$$

$$a = 1.6\angle 40°, \quad b = 1.0\angle 180°, \quad c = 0.9\angle -145°, \quad d = 0.4\angle 148°$$

$$1 \le N \le 5, \overline{\rho} = 0.2N, \qquad 6 \le N \le 10, \overline{\rho} = 5/(10 - N)$$

Fig.(3.3.2) (constant $\overline{\rho}$ circles are given for integer values of N)

Γ_L-plane Γ_{IN}-plane mapping

For the mapping of the Γ_{IN}-plane into the Γ_L-plane, as defined by (3.3.1), we substitute $S_{11}, -\Delta_S, 1$ and $-S_{22}$ for a, b, c and d respectively in (3.3.5) and (3.3.6) with $\overline{\rho} = \rho_{IN}$. Hence for the circles of constant $|\Gamma_{IN}| = \rho_{IN}$, we have

$$\Gamma_{LC\rho_{IN}} = \frac{-S_{11}\Delta_S^* + \rho_{IN}^2 S_{22}^*}{\rho_{IN}^2 |S_{22}|^2 - |\Delta_S|^2}, \tag{3.3.12}$$

$$\rho_{LC\rho_{IN}} = \frac{\rho_{IN}|S_{12}S_{21}|}{\left|\rho_{IN}^2 |S_{22}|^2 - |\Delta_S|^2\right|} \tag{3.3.13}$$

as the centers and radii of the corresponding circles in the Γ_L-plane. In the above equations the subscripts L and C stand for the Γ_L-plane and the 'Circle' respectively.

For $\rho_{IN} = 0$, $\Gamma_{LC\rho_{IN}} = S_{11} / \Delta_S$ and $\rho_{LC\rho_{IN}} = 0$ and for $\rho_{IN} \to \infty$, $\Gamma_{LC\rho_{IN}} \to 1 / S_{22}$ and $\rho_{LC\rho_{IN}} \to 0$. For $\rho_{IN} \to 1$, $\Gamma_{LC\rho_{IN}}$ and $\rho_{LC\rho_{IN}}$ are given as

$$\Gamma_{LC\rho_{IN}} \Big|_{\rho_{IN} \to 1} = \frac{S_{22}^* - S_{11}\Delta_S^*}{|S_{22}|^2 - |\Delta_S|^2}, \tag{3.3.14}$$

$$\rho_{LC\rho_{IN}} \Big|_{\rho_{IN} \to 1} = \frac{|S_{12}S_{21}|}{\left||S_{22}|^2 - |\Delta_S|^2\right|}. \tag{3.3.15}$$

For the radial lines of $T_{IN} = \tan(\tau_{IN})$ in the Γ_{IN}-plane, we have circles in the Γ_L-plane with centers and radii given by

$$\Gamma_{LC\tau_{IN}} = \frac{(\Delta_S^* + S_{11}S_{22}^*)T_{IN} - j(\Delta_S^* - S_{11}S_{22}^*)}{(\Delta_S S_{22}^* + \Delta_S^* S_{22})T_{IN} + j(\Delta_S S_{22}^* - \Delta_S^* S_{22})}, \tag{3.3.16}$$

$$\rho_{LC\tau_{IN}} = \frac{|S_{12}S_{21}|\sqrt{(T_{IN}^2 + 1)}}{\left|(\Delta_S S_{22}^* + \Delta_S^* S_{22})T_{IN} + j(\Delta_S S_{22}^* - \Delta_S^* S_{22})\right|}. \tag{3.3.17}$$

For some applications, it may be desirable to map points in the Γ_L-plane into the Γ_{IN}-plane. In these cases the circles of constant magnitude and lines of constant phase in the Γ_L-plane map into two orthogonal families of circles in the Γ_{IN}-plane.

The centers and radii of circles in the Γ_{IN}-plane for constant ρ_L can be found by substituting S_{11}, -1, Δ_S and $-S_{22}$ for a, b, c and d respectively in (3.3.5) and (3.3.6) with $\overline{\rho} = \rho_L$.

These are given as

$$\Gamma_{INC\rho_L} = \frac{-S_{11} + \rho_L^2 \Delta_S S_{22}^*}{\rho_L^2 |S_{22}|^2 - 1}, \tag{3.3.18}$$

$$\rho_{INC\rho_L} = \frac{\rho_L |S_{12} S_{21}|}{\left| \rho_L^2 |S_{22}|^2 - 1 \right|}. \tag{3.3.19}$$

For $\rho_L = 0$, $\Gamma_{INC\rho_L} = S_{11}$ and $\rho_{INC\rho_L} = 0$ and for $\rho_L \to \infty$, $\Gamma_{INC\rho_L} \to \Delta_S / S_{22}$ and $\rho_{INC\rho_L} \to 0$. For $\rho_L \to 1$, $\Gamma_{INC\rho_L}$ and $\rho_{INC\rho_L}$ are given as

$$\Gamma_{INC\rho_L} \atop \rho_L \to 1 = \frac{-S_{11} + \Delta_S S_{22}^*}{|S_{22}|^2 - 1}, \tag{3.3.20}$$

$$\rho_{INC\rho_L} \atop \rho_L \to 1 = \frac{|S_{12} S_{21}|}{\left| |S_{22}|^2 - 1 \right|}. \tag{3.3.21}$$

The above circle defines the boundary of the region in the Γ_{IN}-plane that is realizable with passive load impedances.

For the radial lines of $T_L = \tan(\tau_L)$ in the Γ_L-plane, we have circles in the Γ_{IN}-plane with centers and radii given by (3.3.10) and (3.3.11) as

$$\Gamma_{INC\tau_L} = \frac{(\Delta_S + S_{11} S_{22}^*) T_L - j(\Delta_S - S_{11} S_{22}^*)}{(S_{22}^* + S_{22}) T_L + j(S_{22}^* - S_{22})}, \tag{3.3.22}$$

$$\rho_{INC\tau_L} = \frac{|S_{12} S_{21}| \sqrt{(T_L^2 + 1)}}{\left| (S_{22}^* + S_{22}) T_L + j(S_{22}^* - S_{22}) \right|}. \tag{3.3.23}$$

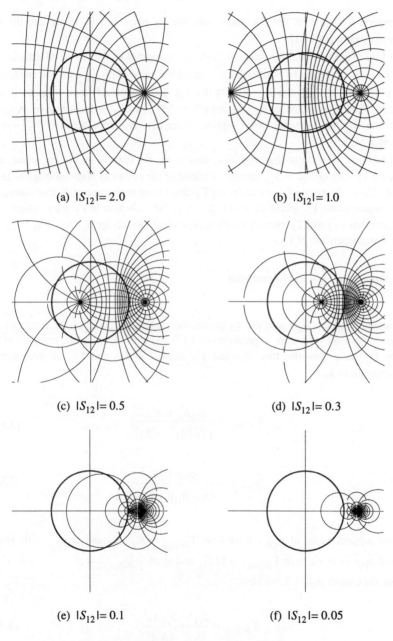

(a) $|S_{12}| = 2.0$

(b) $|S_{12}| = 1.0$

(c) $|S_{12}| = 0.5$

(d) $|S_{12}| = 0.3$

(e) $|S_{12}| = 0.1$

(f) $|S_{12}| = 0.05$

Fig. (3.3.3) Γ_{IN} – plane $S_{11} = 1.4$, $S_{21} = 2.0$ $S_{22} = 0.6$

For typical mappings of the Γ_{IN}-plane into the Γ_L-plane and Γ_L-plane into the Γ_{IN}-plane see example (3.3.1).

In Fig. (3.3.3a) to (3.3.3f) we have given the mappings of the circles of constant ρ_L and T_L into Γ_{IN}-plane, but successively reduced the value of $|S_{12}|$. As it can be seen, the two points for $\rho_L = 0$ and $\rho_L \to \infty$ in the Γ_{IN}-plane approach each other as $|S_{12}|$ is reduced. This is expected as Δ_S / S_{22} approaches S_{11}, as $|S_{12}|$ reduces to zero. A similar effect can also be observed in the Γ_L-plane where S_{11} / Δ_S approaches $1 / S_{22}$, with the reduction of $|S_{12}|$.

In addition we observe that with $|S_{12}|$ decreasing, the regions in the Γ_{IN}-plane, apart from the vicinity of $\Gamma_{IN} = S_{11}$, become depleted of the circles of constant magnitude and phase. Hence to change Γ_{IN} by changing Γ_L (apart from the values of Γ_L that correspond to the region in the Γ_{IN}-plane close to $\Gamma_{IN} = S_{11}$), large changes in Γ_L are required. This indicates that Γ_{IN} and Γ_L become nearly independent. In fact for $S_{12} = 0$, $\Gamma_{IN} = S_{11}$ and is totally independent of Γ_L.

Γ_S-plane and Γ_{OT} -plane mapping

For mapping of the Γ_{OT} into the Γ_S-plane, the centers and radii of the corresponding circles are given by the same expressions (3.3.12) and (3.3.13) as for the mapping of Γ_{IN} in the Γ_L-plane, provided that S_{11} and S_{22} as well as S_{12} and S_{21} are interchanged. These are given as

$$\Gamma_{SC\rho_{OT}} = \frac{-S_{22}\Delta_S^* + \rho_{OT}^2 S_{11}^*}{\rho_{OT}^2 |S_{11}|^2 - |\Delta_S|^2},$$
(3.3.24)

$$\rho_{SC\rho_{OT}} = \frac{\rho_{OT}|S_{12}S_{21}|}{\left|\rho_{OT}^2|S_{11}|^2 - |\Delta_S|^2\right|}.$$
(3.3.25)

For the particular case of $\rho_{OT} = 0$ we have $\Gamma_{SC\rho_{OT}} = S_{22} / \Delta_S$ and $\rho_{SC\rho_{OT}} = 0$. For the case of $\rho_{OT} \to \infty$ we have $\Gamma_{SC\rho_{OT}} \to 1 / S_{11}$ and again $\rho_{SC\rho_{OT}} = 0$.

For the case of $\rho_{OT} = 1$, we have

$$\Gamma_{SC\rho_{OT}} = \frac{S_{11}^* - S_{22}\Delta_S^*}{|S_{11}|^2 - |\Delta_S|^2},$$
$$\rho_{OT} \to 1$$
(3.3.26)

$$\rho_{SC\rho_{OT}} \bigg|_{\rho_{OT} \to 1} = \frac{|S_{12}S_{21}|}{\left| |S_{11}|^2 - |\Delta_S|^2 \right|}. \tag{3.3.27}$$

For the radial lines of $T_{OT} = \tan(\tau_{OT})$ in the Γ_{OT}-plane, we have circles in the Γ_S-plane with centers and radii given by

$$\Gamma_{SC\tau_{OT}} = \frac{(\Delta_S^* + S_{22}S_{11}^*)T_{OT} - j(\Delta_S^* - S_{22}S_{11}^*)}{(\Delta_S S_{11}^* + \Delta_S^* S_{11})T_{OT} + j(\Delta_S S_{11}^* - \Delta_S^* S_{11})}, \tag{3.3.28}$$

$$\rho_{SC\tau_{OT}} = \frac{|S_{12}S_{21}|\sqrt{(T_{OT}^2 + 1)}}{\left| (\Delta_S S_{11}^* + \Delta_S^* S_{11})T_{OT} + j(\Delta_S S_{11}^* - \Delta_S^* S_{11}) \right|}. \tag{3.3.29}$$

For the mapping of the Γ_S-plane into Γ_{OT}-plane, it is again sufficient to interchange S_{12} and S_{21} together with S_{11} and S_{22} in the equations for the mapping of the Γ_L-plane into the Γ_{IN} -plane. Hence we have

$$\Gamma_{OTC\rho_S} = \frac{-S_{22} + \rho_S^2 \Delta_S S_{11}^*}{\rho_S^2 |S_{11}|^2 - 1}, \tag{3.3.30}$$

$$\rho_{OTC\rho_S} = \frac{\rho_S |S_{12}S_{21}|}{\left| \rho_S^2 |S_{11}|^2 - 1 \right|}. \tag{3.3.31}$$

For the particular cases of $\rho_S = 0$ and $\rho_S \to \infty$ we have $\Gamma_{OTC\rho_S} = S_{22}$ and $\Gamma_{OTC\rho_S} = \Delta_S / S_{11}$ respectively and $\rho_{OTC\rho_S} = 0$ for both cases. For the case of $\rho_S = 1$, we have

$$\Gamma_{OTC\rho_S} \bigg|_{\rho_S \to 1} = \frac{-S_{22} + \Delta_S S_{11}^*}{|S_{11}|^2 - 1}, \tag{3.3.32}$$

$$\rho_{OTC\rho_S} \bigg|_{\rho_S \to 1} = \frac{|S_{12}S_{21}|}{\left| |S_{11}|^2 - 1 \right|}. \tag{3.3.33}$$

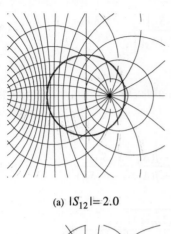

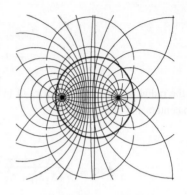

(a) $|S_{12}| = 2.0$

(b) $|S_{12}| = 1.0$

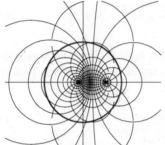

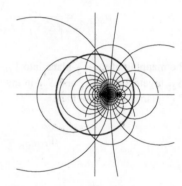

(c) $|S_{12}| = 0.5$

(d) $|S_{12}| = 0.3$

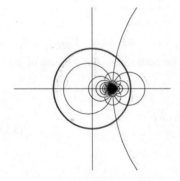

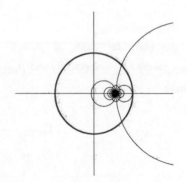

(e) $|S_{12}| = 0.1$

(f) $|S_{12}| = 0.05$

Fig(3.3.4) Γ_{OT} – plane $S_{11} = 1.4, \quad S_{21} = 2.0, \quad S_{22} = 0.6$

For the radial lines of $T_S = \tan(\tau_S)$ in the Γ_S-plane, we have circles in the Γ_{OT}-plane with centers and radii given by

$$\Gamma_{OTC\tau_S} = \frac{(\Delta_S + S_{22}S_{11}^*)T_S - j(\Delta_S - S_{22}S_{11}^*)}{(S_{11}^* + S_{11})T_S + j(S_{11}^* - S_{11})},\tag{3.3.34}$$

$$\rho_{OTC\tau_S} = \frac{|S_{12}S_{21}|\sqrt{(T_S^2 + 1)}}{|(S_{11}^* + S_{11})T_S + j(S_{11}^* - S_{11})|}.\tag{3.3.35}$$

In Fig. (3.3.4a) to (3.3.4f), we have again successively reduced the value of $|S_{12}|$. Similar remarks as for the case of Γ_L and Γ_{IN}-plane mapping are applicable here as $|S_{12}|$ approaches zero. For this case Γ_{OT} becomes equal to S_{22} and is independent of Γ_S when $|S_{12}| = 0$.

Example (3.3.1)

Examine the mapping of the Γ_{IN}-plane into the Γ_L-plane and the Γ_L-plane into the Γ_{IN}-plane for a network of scattering parameters,

$$S_{11} = 1.6\angle 40°, \quad S_{12} = 0.5\angle 24°,$$

$$S_{21} = 3.0\angle 0°, \quad S_{22} = 0.4\angle -32°.$$

Specifically, find the centers and the radii of the circles of constant $\rho_{IN} = 0$, 1 and ∞ in the Γ_L-plane and constant $\rho_L = 0$, 1 and ∞ in the Γ_{IN}-plane.

Solution

With $\Gamma_{IN} = \rho_{IN} \exp(j\tau_{IN})$, the relevant expressions for the centers and radii of circles of constant ρ_{IN} in the Γ_L-plane are given by (3.3.12) and (3.3.13) and for circles of constant τ_{IN} by (3.3.16) and (3.3.17). These circles are plotted below.

In particular for $\rho_{IN} = 0$ or when the input impedance of the network is equal to the reference impedance, the circle reduces to a point in Γ_L – plane at $\Gamma_{LC\rho_{IN}} = S_{11}/\Delta_S$ $= (1.6\angle 40°)/(0.90\angle -145°) = 1.77\angle 185°$. When $\rho_{IN} \to \infty$, again the radius of the circle reduces to zero and its center approaches the point at $\Gamma_{LC\rho} = 1/S_{22} = 1/$ $(0.4\angle -32°) = 2.5\angle 32°$. The above two points, however, both have $|\Gamma_L| > 1$ and hence cannot be realized by passive impedances.

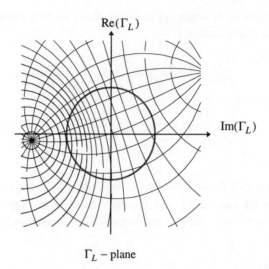

Γ_L – plane

The Γ_L-plane can be divided into two regions with Γ_L producing values of $|\Gamma_{IN}|$ less than or greater than unity. As we shall see in the next chapter this division has great relevance to stability considerations of the network. The boundary of these regions is a circle with the center and radius given by (3.3.14) and (3.3.15) respectively. For this case

$$\Gamma_{LC\rho_{IN}}\Big|_{\rho_{IN} \to 1} = 2.77\angle190.5°,$$

$$\rho_{LC\rho_{IN}}\Big|_{\rho_{IN} \to 1} = 2.29.$$

With $\Gamma_L = \rho_L \exp(j\tau_L)$ the relevant expressions for the centers and radii of circles of constant ρ_L in the Γ_{IN}-plane are given by (3.3.18) and (3.3.19) and for circles of constant τ_L by (3.3.22) and (3.3.23). These circles are plotted in Fig. (3.3.3b). In particular when $\rho_L = 0$ or when the load impedance is equal to the reference impedance, the circle reduces to a point at $\Gamma_{INC\rho} = S_{11} = 1.6\angle40°$. When $\rho_L \to \infty$, again the radius of the circle reduces to zero and its center approaches $\Gamma_{INC\rho} = \Delta_S/S_{22} = 2.26\angle247.3°$.

For passive terminations the input reflection coefficients are confined to a circular region with the center and radius given by (3.3.20) and (3.3.21) as $\Gamma_{INC\rho_L} = 2.29\angle44.9°$ and $\rho_{INC\rho_L} = 1.786$.

$$\text{Re}(\Gamma_{IN})$$

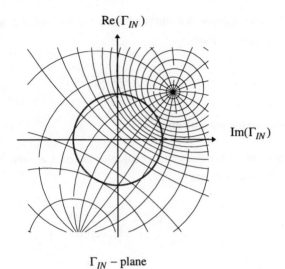

$$\text{Im}(\Gamma_{IN})$$

$$\Gamma_{IN} - \text{plane}$$

3.4 Impedance Matching

For many applications the direct connection of a load to a source of a certain impedance is not desirable. Impedance matching networks can be used to change the impedance of the load, as seen by the source, to any required impedance value. Alternatively for a given network, certain source and load impedances are required to enhance the operation of the network. Specifically, conjugate matching maximizes the transfer of power from the source to the load (see Chapter 5). In this section conjugate matching is defined and matching circuits are discussed.

Conjugate matching

A 2-port network is conjugate matched to the source if $Z_{IN} = Z_S^*$, or equivalently, $\Gamma_{IN} = \Gamma_S^*$. Hence from (3.2.1), for the input impedance to be a conjugate match of the source impedance, the following relation between the source and load reflection coefficients should hold

$$\Gamma_S = \frac{S_{11}^* - \Delta_S^* \Gamma_L^*}{1 - S_{22}^* \Gamma_L^*} \qquad (3.4.1)$$

Similarly a 2-port network is conjugate matched to the load if $Z_{OT} = Z_L^*$ or $\Gamma_{OT} = \Gamma_L^*$. Hence from (3.2.2), we have

$$\Gamma_L = \frac{S_{22}^* - \Delta_S^* \Gamma_S^*}{1 - S_{11}^* \Gamma_S^*}. \tag{3.4.2}$$

Simultaneous conjugate matching

A 2-port network is simultaneously conjugate matched at both ports if $\Gamma_S = \Gamma_{IN}^*$ and $\Gamma_L = \Gamma_{OT}^*$. From (3.4.1) and (3.4.2)

$$\Gamma_{SM} = \frac{S_{11}^* - \Delta_S^* \Gamma_{LM}^*}{1 - S_{22}^* \Gamma_{LM}^*} \tag{3.4.3}$$

and similarly

$$\Gamma_{LM} = \frac{S_{22}^* - \Delta_S^* \Gamma_{SM}^*}{1 - S_{11}^* \Gamma_{SM}^*} \tag{3.4.4}$$

where Γ_{SM} and Γ_{LM} denote, respectively, the required reflection coefficients of the input and output ports for simultaneous conjugate matching.

For simultaneous conjugate matching, conditions (3.4.3) and (3.4.4) should be simultaneously satisfied. Solving for Γ_{SM} in (3.4.4), we have

$$\Gamma_{SM} = \frac{S_{22} - \Gamma_{LM}^*}{\Delta_S - S_{11} \Gamma_{LM}^*}. \tag{3.4.5}$$

Equating the right hand sides of (3.4.3) and (3.4.5) and with some rearrangement, we finally find an equation for Γ_{LM} in the form of

$$C_L \Gamma_{LM}^2 - 2 B_L \Gamma_{LM} + C_L^* = 0 \tag{3.4.6}$$

where B_L and C_L are given as

$$B_L = \frac{1 - |S_{11}|^2 + |S_{22}|^2 - |\Delta_S|^2}{2|S_{12} S_{21}|} \tag{3.4.7}$$

and

$$C_L = \frac{S_{22} - \Delta_S S_{11}^*}{|S_{12} S_{21}|}. \tag{3.4.8}$$

Equation (3.4.6) has a solution in the form of

$$\Gamma_{LM} = \frac{1}{C_L}[B_L \pm \sqrt{(B_L^2 - C_L C_L^*)}]. \tag{3.4.9}$$

For the above equation we can consider the two situations of $B_L^2 - C_L C_L^* < 0$ and $B_L^2 - C_L C_L^* > 0$. For the case of

$$B_L^2 - C_L C_L^* < 0$$

(3.4.9) can be written as

$$\Gamma_{LM} = \frac{1}{C_L}[B_L \pm j\sqrt{(C_L C_L^* - B_L^2)}]$$

and the magnitude of Γ_{LM} as

$$|\Gamma_{LM}| = \sqrt{\left|\frac{1}{C_L}\right|^2 (B_L^2 + |C_L|^2 - B_L^2)} = 1$$

which is physically realizable only for reactive load impedances. From a similar expression for Γ_{SM}, as we shall find shortly, it can be verified that for $B_L^2 - C_L C_L^* < 0$, the source impedance is also purely reactive. Hence for the simultaneous conjugate matching, not leading to the above trivial situation, we require that

$$B_L^2 - C_L C_L^* > 0. \tag{3.4.10}$$

Substituting for B_L and C_L into the above expression, we have

$$\left(1 + |S_{22}|^2 - |S_{11}|^2 - |\Delta_S|^2\right)^2 - 4\left|\left(S_{22} - \Delta_S S_{11}^*\right)\right|^2 > 0$$

and after expansion and rearrangement we require that

$$\left(1 - |S_{22}|^2 - |S_{11}|^2 + |\Delta_S|^2\right)^2 > 4|S_{12} S_{21}|^2.$$

Defining K as

$$K = \frac{1 - |S_{11}|^2 - |S_{22}|^2 + |\Delta_S|^2}{2|S_{12}S_{21}|} \qquad (3.4.11)$$

condition (3.4.10) can simply be written as

$$|K| > 1. \qquad (3.4.12)$$

From (3.4.7), (3.4.8) and (3.4.11) we can write

$$B_L^2 - C_L C_L^* = (K^2 - 1) \qquad (3.4.13)$$

and hence from (3.4.13) and (3.4.9), we finally find two solutions for Γ_{LM} as

$$\Gamma_{LMAX} = \frac{1}{C_L}\left[B_L \mp \sqrt{(K^2 - 1)}\right] \qquad B_L > 0, \qquad (3.4.14a)$$

$$\Gamma_{LMIN} = \frac{1}{C_L}\left[B_L \pm \sqrt{(K^2 - 1)}\right] \qquad B_L < 0. \qquad (3.4.14b)$$

The significance of the designations Γ_{LMAX} and Γ_{LMIN} will be considered in Chapter 5 when we examine the power relations in a 2-port network. The upper signs in (3.4.14) lead to values of $|\Gamma_{LM}| < 1$ required for passive termination while for the lower signs $|\Gamma_{LM}| > 1$ and the load impedance Z_{LM} needs a negative real part.

In a similar manner we can find the reflection coefficient for the source as

$$\Gamma_{SMAX} = \frac{1}{C_S}\left[B_S \mp \sqrt{(K^2 - 1)}\right] \qquad B_S > 0, \qquad (3.4.15a)$$

$$\Gamma_{SMIN} = \frac{1}{C_S}\left[B_S \pm \sqrt{(K^2 - 1)}\right] \qquad B_S < 0 \qquad (3.4.15b)$$

where B_S and C_S are given as

$$B_S = \frac{1 - |S_{22}|^2 + |S_{11}|^2 - |\Delta_S|^2}{2|S_{12}S_{21}|} \qquad (3.4.16)$$

and

$$C_S = \frac{S_{11} - \Delta_S S_{22}^*}{|S_{12}S_{21}|}. \tag{3.4.17}$$

Again in this case, the upper signs in (3.4.15) lead to values of $|\Gamma_{SM}| < 1$ required for passive terminations. For lower signs, however, $|\Gamma_{SM}| > 1$.

A similar condition as (3.4.10), can now be written in this case as

$$B_S^2 - C_S C_S^* > 0 \tag{3.4.18}$$

This, however, leads to the same condition $|K| > 1$ as given by (3.4.12).

Although simultaneous conjugate matching can be obtained when $|K| > 1$ ($K > 1$ or $K < -1$), the condition $K > 1$ is necessary if we consider passive terminations only. Γ_{LM} can then be found from equations (3.4.14a) and (3.4.14b) and Γ_{SM} from (3.4.15a) and (3.4.15b) with upper signs. This point will be made clear in the following chapters.

For the special case of $K = 1$, $B_L = C_L = 0$ and $B_S = C_S = 0$. In this case, as we shall see in Chapter 5, the network is lossless and simultaneous conjugate matching can be achieved for any desired values of either the load or the source impedance. If the load impedance is fixed, the source impedance can be found from (3.4.1). Similarly if the source impedance is fixed, the load impedance can be found from (3.4.2).

The following examples are given for impedance matching. For the last two examples, however, consideration of scattering parameters is not necessary and these are included as examples of transmission line stub matching.

Example (3.4.1)

For the 2-port network of Example (3.2.1) terminated by an impedance $Z_L = 50 + j25$, find the required source impedance for the condition of conjugate matching at the source.

Solution

The scattering parameters of the network were given as

$$S_{11} = 0.6\angle 30°, \quad S_{12} = 0.02\angle 15°,$$

$$S_{21} = 2.0\angle 0, \quad S_{22} = 0.8\angle -45°$$

and the input reflection coefficient from example(3.2.1) as

$$\Gamma_{IN} = 0.518 + j0.311.$$

Hence for the conjugate matching at the source we require,

$$\Gamma_S = \Gamma_{IN}^* = 0.518 - j0.311 \ (Z_S = 96.4 - j94.6).$$

Example (3.4.2)

For the 2-port network of the previous example, if the source impedance is $Z_S = 30 - j45$, calculate the load impedance for the conjugate matching of the output port.

Solution

From Example (3.2.2)
$$\Gamma_{OT} = 0.567 - j0.590$$

and hence for the conjugate matching at the load

$$\Gamma_L = \Gamma_{OT}^* = 0.567 + j0.590 \ (Z_L = 30.8 + j110.2).$$

Example (3.4.3)

For the previous network verify that the simultaneous conjugate matching is possible and find the appropriate source and load impedances.

Solution

For the given scattering parameters

$$\Delta_S = S_{11}S_{22} - S_{12}S_{21} = 0.425 - j0.134$$

giving $K = 2.48 > 1$ and hence the simultaneous matching can be realized.
The parameters B_S and C_S can be found as

$$B_S = \left(1 + |S_{11}|^2 - |S_{22}|^2 - |\Delta_S|^2\right) / \left(2|S_{12}S_{21}|\right) = 6.52$$

and

$$C_S = \left(S_{11} - \Delta_S S_{22}^*\right) / |S_{12}S_{21}| = 5.08 + j3.39.$$

Hence we have

$$\Gamma_{SM} = 0.578 - j0.386 \ (Z_{SM} = 79.0 - j117.9).$$

Similarly

$$B_L = \left(1 + |S_{22}|^2 - |S_{11}|^2 - |\Delta_S|^2\right) / \left(2|S_{12}S_{21}|\right) = 13.52$$

and

$$C_L = \left(S_{22} - \Delta_S S_{11}^*\right) / |S_{12}S_{21}| = 9.63 - j9.21$$

and hence

$$\Gamma_{LM} = 0.610 + j0.583 \ (Z_{LM} = 29.3 + j118.5).$$

Example (3.4.4)

Show that the L-sections of Example(2.3.7) transform the given load impedances Z_L to a 50 ohms impedance, using the scattering parameters of the L-sections as given in the same Example. Verify that $|\Delta_S = 1|$ and $K = 1$ for all networks. The load impedances are given as

(a) $Z_L = 25 + j50$ (b) $Z_L = 40 + j5$ (c) $Z_L = 10 - j30$ (d) $Z_L = 40 - j120$

(e) $Z_L = 5 + j35$ (f) $Z_L = 100$ (g) $Z_L = 50 - j50$ (h) $Z_L = 5 - j35$

The matching properties of the networks can be verified by showing that Γ_{IN} in Equation (3.2.1) is zero. Alternatively we set $\Gamma_{IN} = 0$ in (3.2.1) and verify that $\Gamma_L = S_{11} / \Delta_S$. For the given networks we can make the following table

Network	Δ_S	S_{11} / Δ_S	Γ_L
a	-0.923 + j0.385	0.077 + j0.615	0.077 + j0.615
b	-0.507 - j0.860	-0.108 + j0.062	-0.108 + j0.062
c	1.000 + j0.000	-0.333 - j0.667	-0.333 - j 0.667
d	-0.599 + j0.800	0.600 - j0.533	0.600 - j0.533
e	-0.882 - j0.471	-0.294 + j0.824	-0.294 + j0.824
f	0.000 + j1.000	0.333 - j0.000	0.333 - j0.000
g	0.600 - j0.800	0.200 - j0.400	0.200 - j0.400
h	-0.888 + j0.471	-0.294 - j0.823	-0.294 - j0.823

As it can be seen, $S_{11} / \Delta_S = \Gamma_L$ and hence the above L-sections can be used for the impedance matching.

From the scattering parameters, we can verify that $|S_{11}| = 1 - |S_{21}|$ and $|S_{22}| = 1 - |S_{12}|$. The above table also shows that $|\Delta_S| = 1$ and hence $K = 1$ in all cases, as required for lossless networks (see Sec. 5.5).

Example (3.4.5)

Using the Smith Chart, show that the networks (a) and (e) of the Example (3.4.4) (or Example(2.3.7)), match the load to a source of impedance 50.

Solution

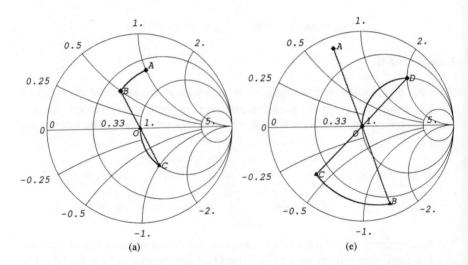

(a)   (e)

(a) Normalizing the load impedance $Z_L = 25 + j50$ by 50, gives $z_L = 0.5 + j1.0$. This normalized impedance can be located at the point A on the Smith Chart (used as an impedance chart). To add a normalized series reactance of $-j / (0.04 \times 50) = -j0.5$, we move point A to point B on the same Chart. Now a normalized parallel susceptance has to be added and this requires the Smith Chart to be used as an admittance chart. To achieve this, point B has to be reflected about the origin of the Chart to point C. This point on the Chart is the normalized admittance of the combination of the load and the series capacitance. Addition of the normalized susceptance $j0.02 / 0.02 = j1.0$, now moves point C to the origin of the Chart. The input impedance of the combined circuit and the load is the required impedance of 50.

(b) In this case, we need to use the Smith Chart initially as an admittance chart. The impedance $Z_L = 5 + j35$ has to be normalized to give $z_L = 0.1 + j0.7$ (point A on the Chart) and A to be reflected about the origin to point B. The reading on the Chart $y_L = 0.2 - j1.4$ at this point is the susceptance of the load. We now add a normalized susceptance of $j0.02 / 0.02 = j1.0$, moving point B to C. The normalized susceptance at this point is $0.2 - j0.4$, giving a normalized impedance of $1.0 + j2.0$ (point D). Addition of a normalized reactance of $-j / (0.01 \times 50)$ $= -j2.0$ moves point D to the origin of the Chart, as required.

Example (3.4.6)

Using a transmission line stub arrangement as shown, find the length of the transmission line section and the short-circuited stub section to conjugate match a load of $Z_L = 50$ to a source of impedance $Z_S = 15.0 - j7.5$. Assume a transmission line characteristic impedance of 50.

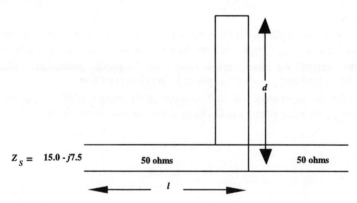

Solution

Normalizing all impedances with respect to the characteristic impedance of the line we have

$$z_L = 1.0,$$

$$z_S = 0.245 - j0.165,$$

$$z_{IN} = z_S^* = 0.245 + j0.165$$

where z_L, z_S and z_{IN} denote the normalized values of Z_L, Z_S and Z_{IN}.

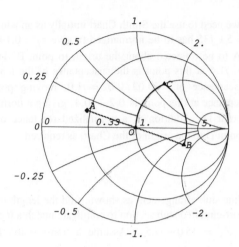

For this problem we make use of the Smith Chart as an admittance chart. The normalized impedances z_{IN} and z_L are located on the Chart as points A and O, and by reflection about the origin we find, respectively, the required normalized admittances $y_{IN} = 2.81 - j1.90$ and $y_L = 1.00$ as points B and O on the Chart.

A section of short-circuited stub of appropriate length adds susceptance to the admittance y_L to move point O to C. Point C should now lie on the circle $|\Gamma_{IN}| = 0.615$, where

$$\Gamma_{IN} = \frac{1 - y_{IN}}{1 + y_{IN}}.$$

The required susceptance can be read on the Chart as $j1.55$ and the length of the short-circuited stub, using the Smith Chart again, as 0.409λ. λ is the wavelength of the propagating wave on the stub line.

The required length of transmission line is found on the Chart as the fraction of wavelength that we have to move clockwise from B to C. This is readily found from the Chart as $(0.278 - 0.177)\lambda = 0.101\lambda$.

Example (3.4.7)

Solve the previous problem, using a different configuration as shown. In this case the lengths of transmission lines are fixed but the characteristic impedances of the line do not need to be equal to the reference impedance 50.

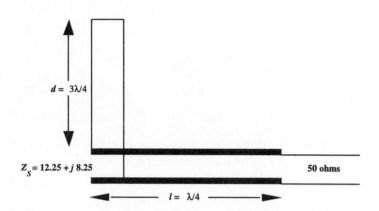

The admittance $Y_{IN} = Y_S^* = 0.056 - j0.038$ and its real part $G_{IN} = 0.056$. This conductance can be achieved by a section of quarterwave transmission line of characteristic impedance

$$Z_{0\,(Trans.\,Line)} = \sqrt{(50)(1/.056)} = 29.9.$$

An open-circuited transmission line has an input susceptance of $jY_0 \tan(2\pi d / \lambda)$. With $d = 3\lambda / 8$, the input susceptance is given simply by $-jY_0$. Hence with required additional susceptance of -0.038, the characteristic impedance of the stub line is given by $Z_{0\,(Stub)} = 1 / 0.038 = 26.3$.

Chapter 4

Stability Considerations of a 2-port Network

Network stability is an important circuit consideration. The term network instability refers to the tendency of a network to oscillate at some frequencies over the electro-magnetic spectrum. Unless a network is to be used as an oscillator, any such oscillations would interfere with the proper functioning of the network. Considering a set of specific source and load impedances, a network is either *stable* or *unstable*, depending on whether any induced oscillations, once the source of initial oscillations is removed, can or cannot be sustained by the network. A 2-port network is *unconditionally stable*, if the network is stable for any combination of the source and load impedances. If a network is not unconditionally stable it is *potentially unstable*. The stability of 2-port networks with given scattering parameters is considered in this Chapter.

4.1 Definition and General Stability Considerations

Condition for oscillation

Consider a 2-port network connected to a source and a load as shown in Fig. (4.1.1).

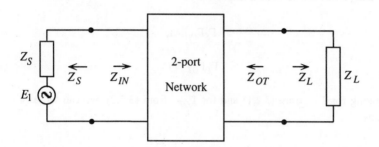

Fig. (4.1.1)

The 2-port network oscillates at its input if

$$Z_S + Z_{IN} = 0 \qquad (4.1.1)$$

and at its output port if

$$Z_L + Z_{OT} = 0. \qquad (4.1.2)$$

In (4.1.1) and (4.1.2), the relations

$$Im(Z_S + Z_{IN}) = 0 \quad \text{and} \quad Im(Z_L + Z_{OT}) = 0$$

set the frequency of oscillations and

$$Re(Z_S + Z_{IN}) = 0 \quad \text{and} \quad Re(Z_L + Z_{OT}) = 0$$

ensure that the oscillations are sustained.
  Substituting for

$$Z_S = \left[\frac{1+\Gamma_S}{1-\Gamma_S}\right] R_0, \qquad Z_{IN} = \left[\frac{1+\Gamma_{IN}}{1-\Gamma_{IN}}\right] R_0$$

and

$$Z_L = \left[\frac{1+\Gamma_L}{1-\Gamma_L}\right]R_0, \qquad Z_{OT} = \left[\frac{1+\Gamma_{OT}}{1-\Gamma_{OT}}\right]R_0$$

(4.1.1) and (4.1.2) can be written respectively as

$$\Gamma_S\Gamma_{IN} = 1, \tag{4.1.3}$$

$$\Gamma_L\Gamma_{OT} = 1. \tag{4.1.4}$$

Substituting for Γ_{IN} from (3.2.1) and for Γ_{OT} from (3.2.2), we can write (4.1.3) and (4.1.4) as

$$\Gamma_S(S_{11} - \Delta_S\Gamma_L) = (1 - \Gamma_L S_{22}) \tag{4.1.5a}$$

and

$$\Gamma_L(S_{22} - \Delta_S\Gamma_S) = (1 - \Gamma_S S_{11}) \tag{4.1.5b}$$

respectively. In (4.1.5a) and (4.1.5b) we have assumed that $S_{12} \neq 0$.

Clearly the conditions (4.1.5a) and (4.1.5b) are identical and hence if $S_{12} \neq 0$, the input and output ports oscillate simultaneously. The frequency of oscillation is also the same.

As S_{12} decreases the time required to reach steady state, and hence to establish a common frequency of oscillation for the input and output circuits, increases. When S_{12} approaches zero, this time approaches infinity and hence a common frequency of oscillations is never achieved.

For the particular case of $S_{12} = 0$, $\Gamma_{IN} = S_{11}$ and $\Gamma_{OT} = S_{22}$ and for the oscillations of the input port we have

$$\Gamma_S S_{11} = 1$$

and for the output port

$$\Gamma_L S_{22} = 1.$$

Hence for the case of $S_{12} = 0$, oscillations of the two ports are independent. It is possible, therefore, that only one port may oscillate or that the frequencies of oscillation of 2 ports are different.

Stability of a network connected to a particular source and load impedance

Consider a network connected to a particular set of source impedance Z_S and load impedance Z_L, as shown in Fig. (4.1.1). If (4.1.1) and (4.1.2) are not satisfied and the network does not oscillate at its operating frequency, it may still remain unstable. The latter situation would occur if the net resistance of either the input or the output circuits is negative. When the net resistance in one circuit is negative, any residual oscillations (such as thermal noise) would grow, until the system nonlinearities limit any further increase in the wave amplitude. As the oscillations reach a steady state or a constant amplitude, the net circuit resistance reduces to zero and the input and output ports oscillate simultaneously, satisfying (4.1.1) and (4.1.2) with a different set of values for Z_{IN} and Z_{OT}. The frequency of oscillations, however, will be different from the operating frequency.

In accordance to the above discussion a 2-port network is stable at its input port and at a particular operating frequency if

$$\text{Re}(Z_S + Z_{IN}) > 0 \tag{4.1.6}$$

and its output port if

$$\text{Re}(Z_L + Z_{OT}) > 0. \tag{4.1.7}$$

If either of the (4.1.6) or (4.1.7) is not satisfied, the system will eventually assume an oscillatory condition as discussed in the previous paragraph.

Stability independent of the source impedance

Although (4.1.6) and (4.1.7) ensure the stability of a network for any particular set of source or load impedances at an operating frequency, it is desirable to find the condition when the input port is stable for *all* values of Z_S. With such a condition satisfied the input circuit does not oscillate with any value of Z_S, and hence, the output circuit is also stable

For circuit stability independent of Z_S, Equation (4.1.6) reduces to

$$\text{Re}(Z_{IN}) > 0 \tag{4.1.8}$$

as $\text{Re}(Z_S)$ is always assumed positive.

Expression (4.1.8) can be written as

$$Z_{IN} + Z_{IN}^* > 0$$

or

$$\left[\frac{1 + \Gamma_{IN}}{1 - \Gamma_{IN}} + \frac{1 + \Gamma_{IN}^*}{1 - \Gamma_{IN}^*} \right] R_0 > 0$$

and hence finally

$$|\Gamma_{IN}| < 1. \tag{4.1.9}$$

If (4.1.9) is satisfied for a particular load impedance, the network is stable for that load impedance, independent of the source impedance.

Stability independent of the load impedance

Similarly it is desirable to find the condition for the circuit to be stable for *all* values of the load impedance Z_L.

For circuit stability independent of Z_L, Equation (4.1.7) reduces to

$$\text{Re}(Z_{OT}) > 0 \tag{4.1.10}$$

as $\text{Re}(Z_L)$ is assumed positive.

Expression (4.1.10) can be written as

$$Z_{OT} + Z_{OT}^* > 0$$

or

$$\left[\frac{1 + \Gamma_{OT}}{1 - \Gamma_{OT}} + \frac{1 + \Gamma_{OT}^*}{1 - \Gamma_{OT}^*} \right] R_0 > 0$$

and hence finally

$$|\Gamma_{OT}| < 1. \tag{4.1.11}$$

If (4.1.11) is satisfied for any particular source impedance, the network is stable for that source impedance, independent of the load impedance.

Unconditional stability or stability for all source or load impedances

An *unconditionally stable* network is a network that is stable for *all* source impedances Z_S and *all* load impedances Z_L.

For unconditional stability, therefore,

$$|\Gamma_{IN}| < 1 \quad \text{for all } \Gamma_L$$

and

$$|\Gamma_{OT}| < 1 \quad \text{for all } \Gamma_S.$$

Unconditional stability is considered in Section (4.3).

4.2 Stability Circles for a Bilateral 2-port Network

In the previous section, we considered the conditions that a network would be stable for all Z_S and also the conditions that a network would be stable for all Z_L. In general, these conditions are satisfied, respectively, for a limited range of load and source impedances. In this section we identify these impedances by specifying the regions of stability in the Γ_L and Γ_S planes.

Substituting for Γ_{IN} from (3.2.1) and Γ_{OT} from (3.2.2), (4.1.9) and (4.1.11) can be written respectively as

$$\frac{|S_{11} - \Gamma_L \Delta_S|}{|1 - \Gamma_L S_{22}|} < 1 \tag{4.2.1}$$

and

$$\frac{|S_{22} - \Gamma_S \Delta_S|}{|1 - \Gamma_S S_{11}|} < 1. \tag{4.2.2}$$

Equating the left and right hand side of equation (4.2.1), we define the points in Γ_L-plane that are on the verge of possible instability. The locus of these points defines the boundary between the stable and potentially unstable regions in Γ_L-plane. Similarly by equating the left and right hand side of equation (4.2.2), we define the points in the Γ_S-plane that are on the verge of possible instability.

The boundary between the stable and potentially unstable regions in the Z_L-plane is hence defined by the expression

$$\frac{|S_{11} - \Gamma_L \Delta_S|}{|1 - \Gamma_L S_{22}|} = 1. \tag{4.2.3}$$

Similarly in the Z_S-plane this boundary is defined by the expression

$$\frac{|S_{22} - \Gamma_S \Delta_S|}{|1 - \Gamma_S S_{11}|} = 1. \tag{4.2.4}$$

Expanding (4.2.3) we have

$$(S_{11} - \Gamma_L \Delta_S)(S_{11}^* - \Gamma_L^* \Delta_S^*) = (1 - \Gamma_L S_{22})(1 - \Gamma_L^* S_{22}^*)$$

or after rearrangement

$$\Gamma_L \Gamma_L^* - \Gamma_{LCS} \Gamma_L^* - \Gamma_{LCS}^* \Gamma_L = \rho_{LCS}^2 - |\Gamma_{LCS}|^2$$

where

$$\Gamma_{LCS} = \frac{S_{22}^* - \Delta_S^* S_{11}}{|S_{22}|^2 - |\Delta_S|^2}, \tag{4.2.5}$$

$$\rho_{LCS} = \frac{|S_{12} S_{21}|}{\left||S_{22}|^2 - |\Delta_S|^2\right|}. \tag{4.2.6}$$

In the complex Γ_L-plane, the loci of points satisfying (4.2.3) is a circle of center at Γ_{LCS} and radius ρ_{LCS}. The circle can be called a Γ_L-plane stability circle. The center and the radius in (4.2.5) and (4.2.6) are obviously identical to those given by (3.3.14) and (3.3.15) of Chapter 3.

To determine whether the inside or the outside of the output stability circle corresponds to a stable condition, we examine the expression (4.2.1) for the simple situation of $\Gamma_L = 0$. It is then clear that the region including the center of the Γ_L-plane where $|\Gamma_L| < 1$, is stable when $|S_{11}| < 1$ and unstable when $|S_{11}| > 1$.

For the case of $S_{12} \to 0$, $\rho_{LCS} \to 0$ and $\Gamma_{LCS} \to 1/S_{22}$ and hence the stability circle reduces to a point at $\Gamma_L = 1/S_{22}$. This point lies inside the $|\Gamma_L| = 1$ circle for $|S_{22}| > 1$ and outside for $|S_{22}| < 1$. The unilateral case is further discussed in section (4.4).

In a similar manner the expression (4.2.4) leads to the Γ_S-plane stability circle. The centers and the radii of the circles are given by

$$\Gamma_{SCS} = \frac{S_{11}^* - \Delta_S^* S_{22}}{|S_{11}|^2 - |\Delta_S|^2} \tag{4.2.7}$$

$$\rho_{SCS} = \frac{|S_{12} S_{21}|}{\left||S_{11}|^2 - |\Delta_S|^2\right|} \tag{4.2.8}$$

Again from expression (4.2.2) at $\Gamma_S = 0$, we conclude that the region including the center of Γ_S -plane is stable when $|S_{22}| < 1$ and potentially unstable when $|S_{22}| > 1$.

For the case of $S_{12} \to 0$, $\rho_{SCS} \to 0$ and $\Gamma_{SCS} \to 1/S_{11}$ and hence the stability circle reduces to a point at $\Gamma_S = 1/S_{11}$.

We again emphasize that for the load impedances for which Γ_L is within the stable region, $\text{Re}(Z_{IN})$ is positive and $|\Gamma_{IN}| < 1$ and the system is stable for any value of Z_S. For the points outside the stable region the system is potentially unstable and could oscillate with a suitable source impedance. Points on the circles are also potentially unstable but the system can only oscillate in the trivial case of the source impedance being purely reactive. Similar remarks can be made for the Γ_S -plane stability circles.

All possible situations that may arise for Γ_L-plane stability circles with respect to the relative values of $|\Gamma_{LCS}|$ and ρ_{LCS}, are shown in Fig. (4.2.1a) and Fig. (4.2.1b). Shaded areas represent the stable region in Fig. (4.2.1a) for $|S_{11}| < 1$ and in Fig. (4.2.1b) for $|S_{11}| > 1$. The stability circles for Γ_S -plane can be drawn similarly.

Dependence of $|\Gamma_{LCS}|$ and ρ_{LCS} on K and B_L

In the last chapter we defined two circuit constants K and B_L, we now relate $|\Gamma_{LCS}|$ and ρ_{LCS} to these constants. This is required for further discussion of the network stability in the next section.

The network constants K and B_L were given by the expressions

$$K = \frac{1 - |S_{11}|^2 - |S_{22}|^2 + |\Delta_S|^2}{2|S_{12}S_{21}|}$$

and

$$B_L = \frac{1 - |S_{11}|^2 + |S_{22}|^2 - |\Delta_S|^2}{2|S_{12}S_{21}|}.$$

From the above expressions we can easily verify the following relations

$$B_L + K = \frac{1 - |S_{11}|^2}{|S_{12}S_{21}|},$$

$$B_L - K = \frac{|S_{22}|^2 - |\Delta_S|^2}{|S_{12}S_{21}|},$$

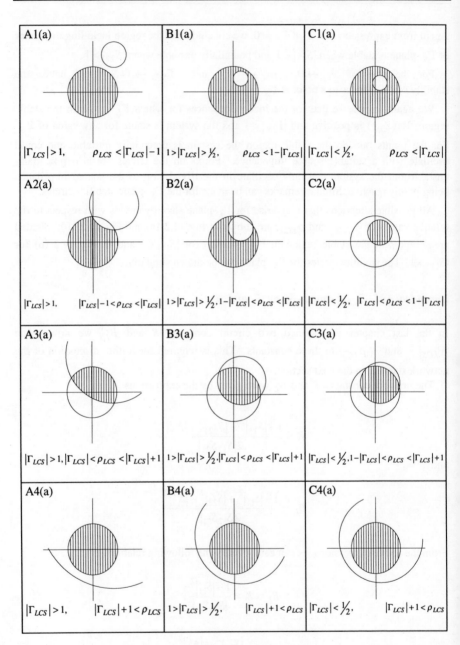

Fig. (4.2.1a): $\left|S_{11}\right| < 1, \quad B_L + K > 0$

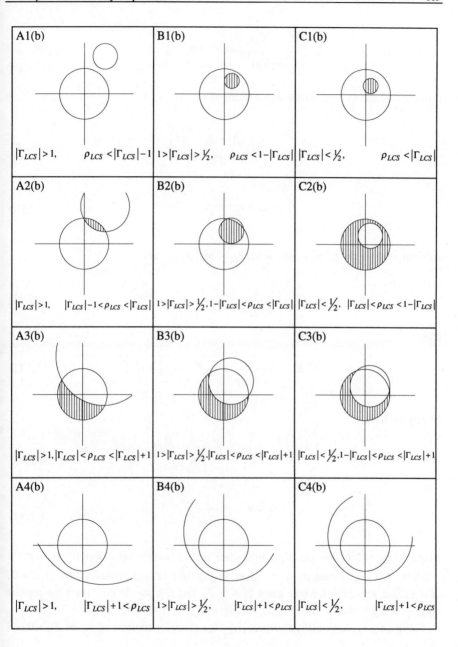

A1(b)	B1(b)	C1(b)
$|\Gamma_{LCS}| > 1,$ $\rho_{LCS} < |\Gamma_{LCS}| - 1$	$1 > |\Gamma_{LCS}| > \frac{1}{2},$ $\rho_{LCS} < 1 - |\Gamma_{LCS}|$	$|\Gamma_{LCS}| < \frac{1}{2},$ $\rho_{LCS} < |\Gamma_{LCS}|$
A2(b)	B2(b)	C2(b)
$|\Gamma_{LCS}| > 1,$ $|\Gamma_{LCS}| - 1 < \rho_{LCS} < |\Gamma_{LCS}|$	$1 > |\Gamma_{LCS}| > \frac{1}{2}, 1 - |\Gamma_{LCS}| < \rho_{LCS} < |\Gamma_{LCS}|$	$|\Gamma_{LCS}| < \frac{1}{2},$ $|\Gamma_{LCS}| < \rho_{LCS} < 1 - |\Gamma_{LCS}|$
A3(b)	B3(b)	C3(b)
$|\Gamma_{LCS}| > 1, |\Gamma_{LCS}| < \rho_{LCS} < |\Gamma_{LCS}| + 1$	$1 > |\Gamma_{LCS}| > \frac{1}{2}, |\Gamma_{LCS}| < \rho_{LCS} < |\Gamma_{LCS}| + 1$	$|\Gamma_{LCS}| < \frac{1}{2}, 1 - |\Gamma_{LCS}| < \rho_{LCS} < |\Gamma_{LCS}| + 1$
A4(b)	B4(b)	C4(b)
$|\Gamma_{LCS}| > 1,$ $|\Gamma_{LCS}| + 1 < \rho_{LCS}$	$1 > |\Gamma_{LCS}| > \frac{1}{2},$ $|\Gamma_{LCS}| + 1 < \rho_{LCS}$	$|\Gamma_{LCS}| < \frac{1}{2},$ $|\Gamma_{LCS}| + 1 < \rho_{LCS}$

Fig. (4.2.1b): $|S_{11}| > 1,$ $B_L + K < 0$

$$\frac{\left|S_{22}^* - \Delta_s^* S_{11}\right|^2}{\left|S_{12} S_{21}\right|^2} = 1 + (B_L^2 - K^2).$$

From (2.4.7) and (4.2.6) and the above relations

$$\left|\Gamma_{LCS}\right| = \frac{\sqrt{[1 + B_L^2 - K^2]}}{\left|B_L - K\right|}, \tag{4.2.9}$$

$$\rho_{LCS} = \frac{1}{\left|B_L - K\right|}. \tag{4.2.10}$$

Alternatively we can relate B_L and K to $\left|\Gamma_{LCS}\right|$ and ρ_{LCS} as

$$K = \pm \frac{1}{2\rho_{LCS}} \left(\left|\Gamma_{LCS}\right|^2 - \rho_{LCS}^2 - 1\right) \tag{4.2.11}$$

and

$$B_L = \pm \frac{1}{2\rho_{LCS}} \left(\left|\Gamma_{LCS}\right|^2 - \rho_{LCS}^2 + 1\right). \tag{4.2.12}$$

Finally we have

$$B_L + K = \pm \frac{\left|\Gamma_{LCS}\right|^2 - \rho_{LCS}^2}{\rho_{LCS}}, \tag{4.2.13}$$

$$B_L - K = \pm \frac{1}{\rho_{LCS}}. \tag{4.2.14}$$

In (4.2.11) to (4.2.14) the plus sign is for when $B_L > K$ and the minus sign for $B_L < K$.

Plots of $\left|\Gamma_{LCS}\right|$ versus ρ_{LCS} are given in Fig. (4.2.2) for constant values of $K = 0$, $K = \pm 1$ and $K = \pm\infty$, for both cases of $K > B_L$ and $K < B_L$. In this figure the regions corresponding to that of Fig. (4.2.1a) and Fig. (4.2.1b) are also indicated.

In Fig. (4.2.3) plots of $\left|\Gamma_{LCS}\right|$ versus ρ_{LCS} are given for constant values of $B_L = 0$ and $B_L = \pm\infty$.

The parameters B_S, K, Γ_{SCS} and ρ_{SCS} are similarly related.

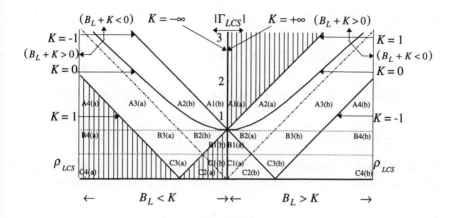

Fig. (4.2.2) Shaded area $K > 1$

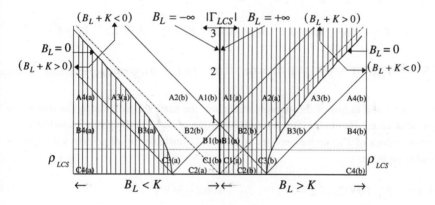

Fig. (4.2.3) Shaded area $B_L > 0$

Example (4.2.1)

For the following 2-port networks with the given scattering parameters, find K and B_L and find the centers and radii of the Γ_L-plane stability circles, as given by expressions (4.2.9) and (4.2.10). Indicate the stable and potentially unstable regions.

Solution

The scattering parameters are given for each case and appropriate constants and relations are derived.

A1(a) $S_{11} = 0.223\angle 153.6°$, $S_{12} = 0.06\angle - 166.4°$,

$S_{21} = 3.0\angle 163.6°$, $S_{22} = 0.65\angle - 56.3°$.

$K = 1.64$, $K > 1$, $B_L = 3.64$, $B_L > 0$ $B_L > K$, $B_L + K > 0$.

$\Gamma_{LCS} = 1.7\angle 60°$, $\rho_{LCS} = 0.5$, $|\Gamma_{LCS}| > 1$, $\rho_{LCS} < |\Gamma_{LCS}| - 1$.

The stability circle is completely outside the $|\Gamma_L| = 1$ circle. With $|S_{11}| < 1$, the origin of the Γ_L-plane is stable and hence the network is stable for all passive load impedances.

A1(b) $S_{11} = 1.40\angle - 121.5°$, $S_{12} = 0.06\angle - 131.5°$,

$S_{21} = 3.0\angle - 161.5°$, $S_{22} = 0.6\angle - 84.2°$.

$K = -1.66$, $K < -1$, $B_L = -3.68$, $B_L < 0$ $B_L < K$, $B_L + K < 0$.

$\Gamma_{LCS} = 1.7\angle 60°$, $\rho_{LCS} = 0.5$, $|\Gamma_{LCS}| > 1$, $\rho_{LCS} < |\Gamma_{LCS}| - 1$.

The center and radius of the stability circle are the same (approximately) as that of the previous case. However, $|S_{11}| > 1$ and the origin of the Γ_L-plane can be unstable (for certain values of Z_S) or the network is potentially unstable for all passive values of the load impedance.

A2(a) $S_{11} = 0.812\angle - 2.9°$, $S_{12} = 0.06\angle - 33.9°$,

$S_{21} = 3.0\angle - 63.9°$, $S_{22} = 0.6\angle - 35.8°$.

$K = 0.44$, $|K| < 1$, $B_L = 1.45$, $B_L > 0$ $B_L > K$, $B_L + K > 0$.

$\Gamma_{LCS} = 1.7\angle 60°$, $\rho_{LCS} = 1.0$, $|\Gamma_{LCS}| > 1$, $|\Gamma_{LCS}| - 1 < \rho_{LCS} < |\Gamma_{LCS}|$.

The stability circle intersects the $|\Gamma_L| = 1$ circle at two points. With $|S_{11}| < 1$, the region in the Γ_L-plane outside the stability circle and containing the origin of the Γ_L-plane is stable. Outside this region the network is potentially unstable for passive load impedances.

A2(b) $S_{11} = 1.158\angle - 44.6°$, $S_{12} = 0.06\angle - 75.9°$,

$S_{21} = 3.0\angle - 105.6°$, $S_{22} = 1.2\angle - 89.8°$.

$K = -0.44$, $|K| < 1$, $B_L = -1.45$, $B_L < 0$ $B_L < K$, $B_L + K < 0$.

$\Gamma_{LCS} = 1.7\angle 60°$, $\rho_{LCS} = 1.0$, $|\Gamma_{LCS}| > 1$, $|\Gamma_{LCS}| - 1 < \rho_{LCS} < |\Gamma_{LCS}|$.

The center and radius of the stability circle are the same (approximately) as that of the previous case. However, $|S_{11}| > 1$ and the origin of the Γ_L-plane is potentially unstable. Hence the region containing the origin is potentially unstable.

A3(a) $S_{11} = 0.965\angle - 123.1°$, $S_{12} = 0.06\angle - 173.5°$,

$S_{21} = 3.0\angle 156.5°$, $S_{22} = 0.62\angle - 139.5°$.

$K = 0.45$, $|K| < 1$, $B_L = -0.07$, $B_L < 0$ $B_L < K$, $B_L + K > 0$.

$\Gamma_{LCS} = 1.7\angle 60°$, $\rho_{LCS} = 1.90$, $|\Gamma_{LCS}| > 1$, $|\Gamma_{LCS}| < \rho_{LCS} < |\Gamma_{LCS}| + 1$.

The stability circle intersects the $|\Gamma_L| = 1$ circle at two points. With $|S_{11}| < 1$, the region inside the stability circle containing the origin of the Γ_L-plane is stable. For other values of Γ_L, the 2-port network is potentially unstable.

A3(b) $S_{11} = 1.034\angle 96.8°$, $S_{12} = 0.06\angle 46.5°$,

$S_{21} = 3.0\angle 16.4°$, $S_{22} = 1.2\angle 25.7°$.

$K = -0.45$, $|K| < 1$, $B_L = 0.07$, $B_L > 0$ $B_L > K$, $B_L + K < 0$.

$\Gamma_{LCS} = 1.7\angle 60°$, $\rho_{LCS} = 1.9$, $|\Gamma_{LCS}| > 1$, $|\Gamma_{LCS}| < \rho_{LCS} < |\Gamma_{LCS}| + 1$.

The center and radius of the stability circle are the same as that of the previous case. However, $|S_{11}| > 1$ and the origin of the Γ_L-plane is potentially unstable. The region outside the stability circle is stable.

A4(a) $S_{11} = 0.796 \angle 81.1°$, $S_{12} = 0.06 \angle 75.4°$,

$S_{21} = 3.0 \angle 45.4°$, $S_{22} = 0.44 \angle -178.9°$.

$K = 1.18$, $K > 1$, $B_L = 0.85$, $B_L > 0$ $B_L < K$, $B_L + K > 0$.

$\Gamma_{LCS} = 1.7 \angle 60°$, $\rho_{LCS} = 3.0$, $|\Gamma_{LCS}| > 1$, $\rho_{LCS} > |\Gamma_{LCS}| + 1$.

The stability circle completely contains the $|\Gamma_L| = 1$ circle. With $|S_{11}| < 1$, the origin and hence all regions of the Γ_L-plane inside the $|\Gamma_L| = 1$ circle are stable

A4(b) $S_{11} = 1.169 \angle 52.3°$, $S_{12} = 0.06 \angle 46.6°$,

$S_{21} = 3.0 \angle 16.6°$, $S_{22} = 0.55 \angle 39.5°$.

$K = -1.18$, $K < -1$, $B_L = -0.85$, $B_L < 0$ $B_L > K$, $B_L + K < 0$.

$\Gamma_{LCS} = 1.7 \angle 60°$, $\rho_{LCS} = 3.0$, $|\Gamma_{LCS}| > 1$, $\rho_{LCS} > |\Gamma_{LCS}| + 1$.

The center and radius of the stability circle are the same as that of the previous case. With $|S_{11}| > 1$, however, the 2-port network is potentially unstable for all passive load impedances.

B1(a) $S_{11} = 0.915 \angle -72.6°$, $S_{12} = 0.06 \angle -60.6°$,

$S_{21} = 3.0 \angle -90.6°$, $S_{22} = 1.3 \angle -45.5°$.

$K = -1.22$, $K < -1$, $B_L = 2.12$, $B_L > 0$ $B_L > K$, $B_L + K > 0$.

$\Gamma_{LCS} = 0.6 \angle 60°$, $\rho_{LCS} = 0.3$, $\frac{1}{2} < |\Gamma_{LCS}| < 1$, $0 < \rho_{LCS} < 1 - |\Gamma_{LCS}|$.

The stability circle is completely inside the circle $|\Gamma_L| = 1$. The circle, however, does not contain the origin of the Γ_L-plane, hence with $|S_{11}| < 1$, the inside of the stability circle is potentially unstable.

B1(b) $S_{11} = 1.078\angle 176.1°,$ $\qquad$ $S_{12} = 0.06\angle 174.1°,$

$S_{21} = 3.0\angle 144.1°,$ $\qquad$ $S_{22} = 1.2\angle -82°.$

$K = 1.22,$ $\quad$ $K > 1,$ $\quad$ $B_L = -2.12,$ $\quad$ $B_L < 0$ $\quad$ $B_L < K,$ $\quad$ $B_L + K < 0.$

$\Gamma_{LCS} = 0.6\angle 60°,$ $\quad$ $\rho_{LCS} = 0.3,$ $\quad$ $\dfrac{1}{2} < |\Gamma_{LCS}| < 1,$ $\quad$ $0 < \rho_{LCS} < 1 - |\Gamma_{LCS}|.$

The situation is similar to that of the previous case. With $|S_{11}| > 1$, however, the stable and potentially unstable regions are interchanged.

B2(a) $S_{11} = 0.98\angle 144.3°,$ $\qquad$ $S_{12} = 0.06\angle -125.5°,$

$S_{21} = 3.0\angle -155.5°,$ $\qquad$ $S_{22} = 1.1\angle -37.9°.$

$K = -0.89,$ $\quad$ $|K| < 1,$ $\quad$ $B_L = 1.11,$ $\quad$ $B_L > 0$ $\quad$ $B_L > K,$ $\quad$ $B_L + K > 0.$

$\Gamma_{LCS} = 0.60\angle 60°$ $\quad$ $\rho_{LCS} = 0.50$ $\quad$ $\dfrac{1}{2} < |\Gamma_{LCS}| < 1,$ $\quad$ $1 - |\Gamma_{LCS}| < \rho_{LCS} < |\Gamma_{LCS}|.$

The stability circle intersects the $|\Gamma_L| = 1$ circle at two points. With $|S_{11}| < 1$, as the origin of Γ_L-plane is outside the stability circle, the region inside the stability circle is potentially unstable. The other parts inside the $|\Gamma_L| = 1$ and outside the stability circles are stable.

B2(b) $S_{11} = 1.020\angle -20.4°,$ $\qquad$ $S_{12} = 0.06\angle 69.8°,$

$S_{21} = 3.0\angle 39.8°,$ $\qquad$ $S_{22} = 1.2\angle -102.1°.$

$K = 0.89,$ $\quad$ $|K| < 1,$ $\quad$ $B_L = -1.11,$ $\quad$ $B_L < 0$ $\quad$ $B_L < K,$ $\quad$ $B_L + K < 0.$

$\Gamma_{LCS} = 0.60\angle 60°, \rho_{LCS} = 0.50, \dfrac{1}{2} < |\Gamma_{LCS}| < 1,$ $\quad$ $1 - |\Gamma_{LCS}| < \rho_{LCS} < |\Gamma_{LCS}|.$

This is similar to the previous situation but with $|S_{11}| > 1$, the stable and potentially unstable regions are interchanged.

B3(a) $S_{11} = 0.954\angle - 157.8°$, $S_{12} = 0.06\angle 166.6°$,

$S_{21} = 3.0\angle 136.6°$, $S_{22} = 0.78\angle - 112.8°$,

$K = 0.81$, $|K| < 1$, $B_L = -0.31$, $B_L < 0$ $B_L < K$, $B_L + K > 0$.

$\Gamma_{LCS} = 0.60\angle 60°$, $\rho_{LCS} = 0.9$, $\dfrac{1}{2} < |\Gamma_{LCS}| < 1$, $|\Gamma_{LCS}| < \rho_{LCS} < |\Gamma_{LCS}|$.

The stability circle intersects the $|\Gamma_L| = 1$ circle at two points. The stability circle contains the origin of the Γ_L-plane and $|S_{11}| < 1$, hence the inside of the stability circle is stable. For load impedances having Γ_L outside the stability circle, the 2-port network is potentially unstable.

B3(b) $S_{11} = 1.044\angle 85.6°$, $S_{12} = 0.06\angle 50.0°$,

$S_{21} = 3.0\angle 20.0°$, $S_{22} = 1.47\angle 23.8°$.

$K = -0.81$, $|K| < 1$, $B_L = 0.31$, $B_L > 0$ $B_L > K$, $B_L + K < 0$.

$\Gamma_{LCS} = 0.60\angle 60°$, $\rho_{LCS} = 0.9$, $\dfrac{1}{2} < |\Gamma_{LCS}| < 1$ $|\Gamma_{LCS}| < \rho_{LCS} < |\Gamma_{LCS}| + 1$.

The position of the stability circle is similar to the previous case. With $|S_{11}| > 1$, however, the stable and potentially unstable regions are interchanged.

B4(a) $S_{11} = 0.832\angle - 40.4°$, $S_{12} = 0.06\angle - 30.3°$,

$S_{21} = 3.0\angle - 60.3°$, $S_{22} = 0.67\angle 127.1°$.

$K = 1.12$, $K > 1$, $B_L = 0.59$, $B_L > 0$ $B_L < K$, $B_L + K > 0$.

$\Gamma_{LCS} = 0.60\angle 60°$, $\rho_{LCS} = 1.9$, $\dfrac{1}{2} < |\Gamma_{LCS}| < 1$, $\rho_{LCS} > |\Gamma_{LCS}| + 1$.

The stability circle contains the whole $|\Gamma_L| = 1$ circle and with $|S_{11}| < 1$, the 2-port network is stable for all load impedances.

B4(b)　　$S_{11} = 1.144\angle 84.5°$,　　　　　$S_{12} = 0.06\angle 94.6°$,

　　　　$S_{21} = 3.0\angle 64.6°$,　　　　　$S_{22} = 0.808\angle 84.1°$.

　　　　$K = -1.12$,　　$K < -1$,　　$B_L = -0.60$,　　$B_L < 0$　　$B_L > K$,　　$B_L + K < 0$.

　　　　$\Gamma_{LCS} = 0.61\angle 60°$,　　$\rho_{LCS} = 1.9$,　　$\dfrac{1}{2} < |\Gamma_{LCS}| < 1$,　　$\rho_{LCS} > |\Gamma_{LCS}| + 1$.

As in the previous case the stability circle contains the whole $|\Gamma_L| = 1$ circle and with $|S_{11}| > 1$ the 2-port network is potentially unstable for all load impedances.

C1(a)　　$S_{11} = 0.979\angle -57.9°$,　　　　　$S_{12} = 0.06\angle -46.9°$,

　　　　$S_{21} = 3.0\angle -76.9°$,　　　　　$S_{22} = 1.6\angle -44.5°$.

　　　　$K = -1.55$,　　$K < -1$,　　$B_L = 1.78$,　　$B_L > 0$　　$B_L > K$,　　$B_L + K > 0$.

　　　　$\Gamma_{LCS} = 0.4\angle 60°$,　　$\rho_{LCS} = 0.3$,　　$|\Gamma_{LCS}| < \dfrac{1}{2}$,　　$0 < \rho_{LCS} < |\Gamma_{LCS}|$.

The stability circle has no intersection with the $|\Gamma_L| = 1$ circle and, furthermore, it lies totally inside the latter. With $|S_{11}| < 1$ and the origin of the Γ_L-plane outside the stability circle, the inside of the stability circle is everywhere potentially unstable. For other values of Γ_L inside the circle $|\Gamma_L| = 1$ and outside the stability circle the network is stable.

C1(b)　　$S_{11} = 1.020\angle 142.0°$,　　　　　$S_{12} = 0.06\angle 151.0°$,

　　　　$S_{21} = 3.0\angle 121.0°$,　　　　　$S_{22} = 1.6\angle -88.5°$.

　　　　$K = 1.54$,　　$K > 1$,　　$B_L = -1.77$,　　$B_L < 0$　　$B_L < K$,　　$B_L + K < 0$.

　　　　$\Gamma_{LCS} = 0.4\angle 60°$,　　$\rho_{LCS} = 0.3$,　　$|\Gamma_{LCS}| < \dfrac{1}{2}$,　　$0 < \rho_{LCS} < |\Gamma_{LCS}|$.

The center and radius of the stability circle are similar to that of the previous case. As $|S_{11}| > 1$ at the origin of the Γ_L-plane, the inside of the stability circle is stable and the outside is potentially unstable.

C2(a) $S_{11} = 0.984\angle -5.4°$, $S_{12} = 0.06\angle 69.9°$,

$S_{21} = 3.0\angle 39.9°$, $S_{22} = 1.1\angle -84.4°$,

$K = 1.09$, $K > 1$, $B_L = -0.91$, $B_L < 0$ $B_L < K$, $B_L + K > 0$.

$\Gamma_{LCS} = 0.4\angle 60°$, $\rho_{LCS} = 0.5$, $|\Gamma_{LCS}| < \dfrac{1}{2}$, $|\Gamma_{LCS}| < \rho_{LCS} < 1 - |\Gamma_{LCS}|$.

In this case $|K| > 1$ and there is no intersection between the stability circle and the $|\Gamma_L| = 1$ circle. With $|S_{11}| < 1$ and the stability circle containing the origin of the Γ_L-plane, the stable region is inside the stability circle.

C2(b) $S_{11} = 1.016\angle -174.9°$, $S_{12} = 0.06\angle -99.6°$,

$S_{21} = 3.0\angle -129.6°$, $S_{22} = 1.3\angle -34.4°$.

$K = -1.09$, $K < -1$, $B_L = 0.91$, $B_L > 0$ $B_L > K$, $B_L + K < 0$.

$\Gamma_{LCS} = 0.4\angle 60°$, $\rho_{LCS} = 0.5$, $|\Gamma_{LCS}| < \dfrac{1}{2}$, $|\Gamma_{LCS}| < \rho_{LCS} < 1 - |\Gamma_{LCS}|$.

With Γ_{LCS} and ρ_{LCS} the same as in the previous case, the stability circle is identical. With $|S_{11}| > 1$, however, the inside of the stability circle is potentially unstable and the outside is stable.

C3(a) $S_{11} = 0.944\angle -55.4°$, $S_{12} = 0.06\angle 40.4°$,

$S_{21} = 3.0\angle 10.4°$, $S_{22} = 0.79\angle -88.3°$.

$K = 0.93$, $|K| < 1$, $B_L = -0.32$, $B_L < 0$ $B_L < K$, $B_L + K > 0$.

$\Gamma_{LCS} = 0.4\angle 60°$, $\rho_{LCS} = 0.8$, $|\Gamma_{LCS}| < \dfrac{1}{2}$, $1 - |\Gamma_{LCS}| < \rho_{LCS} < |\Gamma_{LCS}| + 1$.

As $|K| < 1$, the stability circle intersects the $|\Gamma_L| = 1$ circle at two points. The stability circle encircles the origin of the Γ_L-plane and with $|S_{11}| < 1$, the inside of the stability circle is stable. The other points inside the Γ_L-plane and outside the stability circle are potentially unstable.

C3(b) $S_{11} = 1.053\angle 94.8°$, $S_{12} = 0.06\angle 59.8°$,

$S_{21} = 3.0\angle 29.8°$, $S_{22} = 1.43\angle 22.8°$.

$K = -0.92$, $|K| < 1$, $B_L = 0.32$, $B_L > 0$, $B_L > K$, $B_L + K < 0$.

$\Gamma_{LCS} = 0.4\angle 60°, \rho_{LCS} = 0.8$ $,|\Gamma_{LCS}| < \dfrac{1}{2}$, $1 - |\Gamma_{LCS}| < \rho_{LCS} < |\Gamma_{LCS}| + 1$.

The stability circle is similar to that of the previous case. With $|S_{11}| > 1$ inside the stability circle is potentially unstable and outside stable.

C4(a) $S_{11} = 0.843\angle 121.7°$, $S_{12} = 0.06\angle 63.7°$,

$S_{21} = 3.0\angle 33.7°$, $S_{22} = 0.65\angle 148.1°$.

$K = 1.10$, $K > 1$, $B_L = 0.51$, $B_L > 0$ $B_L < K$, $B_L + K > 0$.

$\Gamma_{LCS} = 0.4\angle 60°$, $\rho_{LCS} = 1.70$, $|\Gamma_{LCS}| < \dfrac{1}{2}$, $\rho_{LCS} > |\Gamma_{LCS}| + 1$.

The stability circle encircles the $|\Gamma_L| = 1$ circle and there is no intersection between the two. With $|S_{11}| < 1$, the 2-port network is stable for all passive load impedances.

C4(b) $S_{11} = 1.1355\angle - 159.2°$, $S_{12} = 0.06\angle 142.8°$,

$S_{21} = 3.0\angle 112.8°$, $S_{22} = 0.78\angle 64.6°$.

$K = -1.10$, $K < -1$, $B_L = -0.51$, $B_L < 0$ $B_L > K$, $B_L + K < 0$.

$\Gamma_{LCS} = 0.4\angle 60°$, $\rho_{LCS} = 1.70$, $|\Gamma_{LCS}| < \dfrac{1}{2}$, $\rho_{LCS} > |\Gamma_{LCS}| + 1$.

The stability circle has the same center and radius as in the previous case (approximately). With $|S_{11}| > 1$, the 2-port network is potentially unstable for all passive values of the load impedance.

4.3 Unconditional Stability

A network is said to be unconditionally stable if the stability condition $|\Gamma_{IN}| < 1$ is satisfied for all values of Γ_L and the stability condition $|\Gamma_{OT}| < 1$ is satisfied for all values of Γ_S. For unconditional stability, therefore, the stability circle in the Γ_L-plane should lie completely outside the circle $|\Gamma_L| = 1$. Similarly the stability circle in the Γ_S-plane should lie totally outside the circle $|\Gamma_S| = 1$. In addition $|S_{11}|$ should be less than unity to ensure that the origin of the Γ_L-plane is stable. $|S_{22}|$ should also be less than unity to ensure that the center of the Γ_S-plane is stable.

We should realize, however, that as unconditional stability is the network stability for all Γ_L and Γ_S, the conditions for unconditional stability relating to the Γ_L-plane stability circle and to that of the Γ_S-plane stability circle are identical. It is sufficient, therefore, to derive these conditions by either considering the Γ_L-plane stability circle or Γ_S-plane stability circle.

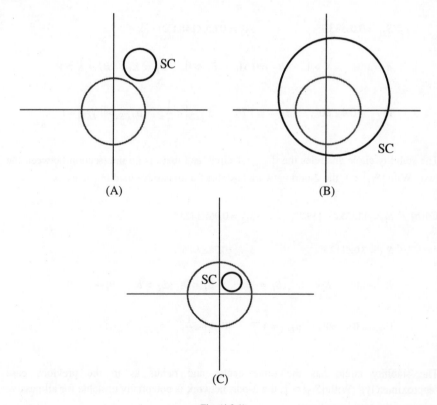

(A)

(B)

(C)

Fig. (4.3.1)

Considering the Γ_L-plane stability circles as in Fig. (4.3.1), the unconditionally stable situations are shown in cases (A) and (B), with the stability circle completely outside the $|\Gamma_L| = 1$ circle. The case (C), however, is potentially unstable. For unconditional stability, therefore, the stability circle and the circle $|\Gamma_L| = 1$ have no intersection in the Γ_L-plane. In addition any small increase in the radius of the latter circle should bring the two circles closer as in the case of (A) and (B) and not farther apart as in the case (C). In the latter case the stability circle is completely inside the $|\Gamma_L| = 1$ circle.

To investigate the above conditions, we examine the intersection of the stability circle with a circle defined by the expression

$$\Gamma_\Lambda = \rho_\Lambda \exp(j\phi_\Lambda)$$

where ρ_Λ is the radius of the circle and ϕ_Λ varies from 0 to 2π.

To find the intersection of the above circle with the stability circle, we substitute Γ_Λ for Γ_L in the equation for the stability circle. This gives

$$\Gamma_\Lambda \Gamma_\Lambda^* - \frac{S_{22}^* - \Delta_S^* S_{11}}{|S_{22}|^2 - |\Delta_S|^2} \Gamma_\Lambda^* - \frac{S_{22} - \Delta_S S_{11}^*}{|S_{22}|^2 - |\Delta_S|^2} \Gamma_\Lambda = \frac{|S_{11}|^2 - 1}{|S_{22}|^2 - |\Delta_S|^2}.$$

With

$$S_{22}^* - \Delta_S^* S_{11} = |S_{22}^* - \Delta_S^* S_{11}| \exp(j\phi_{LCS}),$$

the above equation reduces to

$$\xi(\rho_\Lambda) = \cos(\phi_\Lambda - \phi_{LCS}) = \frac{1 - |S_{11}|^2 + \rho_\Lambda^2 (|S_{22}|^2 - |\Delta_S|^2)}{2\rho_\Lambda |S_{22} - \Delta_S S_{11}^*|}. \tag{4.3.1}$$

For the circle $|\Gamma_\Lambda| = 1$, $\rho_\Lambda = 1$ and we have

$$\xi(1) = \cos(\phi_\Lambda - \phi_{LCS}) = \frac{1 - |S_{11}|^2 + |S_{22}|^2 - |\Delta_S|^2}{2|S_{22} - \Delta_S S_{11}^*|}. \tag{4.3.2}$$

Hence for the stability circle and the circle $|\Gamma_\Lambda| = 1$ to have no intersection, we require that (4.3.2) has no real solution or the square of the right hand side of (4.3.2) to be greater than unity. For unconditional stability, therefore, we have $|\xi(1)| > 1$ or

$$\xi^2(1) > 1$$

and hence

$$(1-|S_{11}|^2+|S_{22}|^2-|\Delta_S|^2)^2 > 4|S_{22}-\Delta_S S_{11}^*|^2.$$

Substituting for

$$|S_{22}-\Delta_S S_{11}^*|^2 = |S_{12}S_{21}|^2+(1-|S_{11}|^2)(|S_{22}|^2-|\Delta_S|^2),$$

we have

$$(1-|S_{11}|^2-|S_{22}|^2+|\Delta_S|^2)^2 > 4|S_{12}S_{21}|^2$$

as a necessary condition for unconditional stability.

With K defined previously as

$$K = \frac{1-|S_{11}|^2-|S_{22}|^2+|\Delta_S|^2}{2|S_{12}S_{21}|},$$

we can write the above condition simply as

$$|K| > 1.$$

Now the derivative of $\xi^2(\rho_\Lambda)$ with respect to ρ_Λ at $\rho_\Lambda = 1$ is given by

$$\left.\frac{d\xi^2(\rho_\Lambda)}{d\rho_\Lambda}\right|_{\rho_\Lambda=1} = 2\xi(\rho_\Lambda)\left.\frac{d\xi(\rho_\Lambda)}{d\rho_\Lambda}\right|_{\rho_\Lambda=1}$$

$$= -\frac{K|S_{12}S_{21}|(1-|S_{11}|^2+|S_{22}|^2-|\Delta_S|^2)}{|S_{22}-\Delta_S S_{11}^*|^2}.$$

With $K > 1$, the derivative of $\xi^2(\rho_\Lambda)$ with respect to ρ_Λ is less than zero, if

$$1-|S_{11}|^2+|S_{22}|^2-|\Delta_S|^2 > 0$$

or

$$B_L > 0$$

where

$$B_L = \frac{1 - |S_{11}|^2 + |S_{22}|^2 - |\Delta_S|^2}{2|S_{12}S_{21}|}$$

as defined before.

With $K < -1$, the derivative of $\xi^2(\rho_\Lambda)$ with respect to ρ_Λ is less than zero, if

$$B_L < 0.$$

We can now conclude that with ($K > 1$ and $B_L > 0$), or ($K < -1$ and $B_L < 0$), the derivative of $\xi^2(\rho_\Lambda)$ with respect to ρ_Λ is negative, and hence any increase in ρ_Λ from $\rho_\Lambda = 1$, decreases the magnitude of the right hand side of (4.3.2) and brings the stability circle and the circle $\Gamma_\Lambda = 1$ closer. This corresponds to cases (A) and (B). Conversely with ($K > 1$ and $B_L < 0$), or ($K < -1$ and $B_L > 0$), any increase in ρ_Λ, moves the two circles further apart as in case (C).

The conditions ($K < -1$ and $B_L < 0$), however, give $B_L + K < 0$ or $|S_{11}| > 1$ and the network is not unconditionally stable. In contrast, the condition ($K > 1$ and $B_L > 0$), corresponds to the case $B_L + K > 0$ or $|S_{11}| < 1$, with the network unconditionally stable. Hence the necessary and sufficient condition for unconditional stability can be written as

$$K > 1 \tag{4.3.3}$$

$$B_L > 0. \tag{4.3.4}$$

Similarly the Γ_S-plane stability circle is completely outside the $|\Gamma_S| = 1$ circle and $K + B_S > 0$ if

$$K > 1 \tag{4.3.5}$$

$$B_S > 0. \tag{4.3.6}$$

where

$$B_S = \frac{1 + |S_{11}|^2 - |S_{22}|^2 - |\Delta_S|^2}{2|S_{12}S_{21}|}.$$

The above conditions for unconditional stability can be verified by considering Figs. (4.2.1a) and (4.2.1b), together with the plots of Figs. (4.2.2) and (4.2.3). Clearly only in situations A1(a), A4(a), B4(a) and C4(a) of Fig. (4.2.1a), the network is unconditionally stable. For these situations we have

$$|\Gamma_{LCS}| > 1, \quad \rho_{LCS} < |\Gamma_{LCS}| - 1 \quad : \text{case A1(a)},$$

$$|\Gamma_{LCS}| > 1, \quad \rho_{LCS} > 1 + |\Gamma_{LCS}| \quad : \text{cases A4(a), B4(a) and C4(a)}.$$

For the above and only for the above situations, as seen in Fig. (4.2.2) and Fig. (4.2.3), both conditions of $K > 1$ and $B_L > 0$, are satisfied.

If a bilateral network is stable at its input port for a particular value of Γ_S and Γ_L, it is also stable at its output port. Consequently if a network is stable for all values of Γ_S and Γ_L at its input port, it is also stable for all values of Γ_S and Γ_L at its output port. Hence the conditions for the unconditional stability for both ports are identical and when the network is unconditionally stable, (4.3.4) and (4.3.6) are simultaneously satisfied.

Writing the conditions (4.3.4) and (4.3.6) as

$$|\Delta_S|^2 < 1 - |S_{11}|^2 + |S_{22}|^2, \tag{4.3.7}$$

$$|\Delta_S|^2 < 1 + |S_{11}|^2 - |S_{22}|^2 \tag{4.3.8}$$

it is clear that for unconditional stability, both (4.3.7) and (4.3.8) are simultaneously satisfied. Adding the left and write hand sides we have

$$|\Delta_S| < 1. \tag{4.3.9}$$

Conversely with (4.3.9) satisfied, one of the relations (4.3.7) or (4.3.8) has obviously be true, and hence the other relation has to be true also. We therefore conclude that condition (4.3.9) can replace both (4.3.4) and (4.3.6). An alternative set of necessary and sufficient conditions for unconditional stability is hence

$$K > 1, \tag{4.3.10}$$

$$|\Delta_S| < 1. \tag{4.3.11}$$

Example (4.3.1)

For all networks of problem (4.2.1) calculate K, B_L, B_S, and $|\Delta_S|$ and determine the stability condition of the networks.

Solution

Tabulated values of K, B_L, B_S, and $|\Delta_S|$ are given. Clearly the networks designated by A1(a), A4(a), B4(a) and C4(a) are the only unconditionally stable networks.

Network	K	B_L	B_S	$|\Delta_S|$	Stability condition
A1(a)	1.66	3.64	1.57	0.25	Unconditionally stable
A2(a)	0.44	1.45	3.11	0.42	Potentially unstable *
A3(a)	0.45	-0.07	2.97	0.69	Potentially unstable*
A4(a)	1.18	0.85	3.29	0.50	Unconditionally stable
B1(a)	-1.22	2.12	-2.62	1.04	Potentially unstable*
B2(a)	-0.89	1.11	-0.28	0.92	Potentially unstable*
B3(a)	0.81	-0.31	1.37	0.90	Potentially unstable*
B4(a)	1.12	0.59	1.94	0.74	Unconditionally stable
C1(a)	-1.55	1.78	-7.12	1.40	Potentially unstable*
C2(a)	1.10	-0.91	-2.26	1.25	Potentially unstable*
C3(a)	0.93	-0.32	1.16	0.92	Potentially unstable*
C4(a)	1.10	0.51	2.11	0.73	Unconditionally stable
A1(b)	-1.66	-3.68	5.21	0.85	Potentially unstable
A2(b)	-0.44	-1.45	-2.01	1.27	Potentially unstable *
A3(b)	-0.45	0.07	-1.99	1.16	Potentially unstable*
A4(b)	-1.18	-0.85	5.06	0.49	Potentially unstable
B1(b)	1.22	-2.12	-3.66	1.43	Potentially unstable*
B2(b)	0.89	-1.11	-3.34	1.34	Potentially unstable*
B3(b)	-0.81	0.31	-5.64	1.40	Potentially unstable*
B4(b)	-1.12	-0.60	3.05	0.75	Potentially unstable
C1(b)	1.54	-1.77	-10.2	1.78	Potentially unstable*
C2(b)	-1.09	0.91	-2.74	1.15	Potentially unstable*
C3(b)	-0.92	0.32	-4.88	1.35	Potentially unstable*
C4(b)	-1.10	-0.51	3.27	0.71	Potentially unstable

For the cases shown by an asterisk, stability can be achieved by all source impedances and a set of load impedances or by all load impedances and a set of source impedances.

4.4 The Stability of Unilateral 2-port Networks

As discussed in the previous chapter, when the scattering parameter S_{12} of a 2-port network is negligibly small, the 2-port network is said to be unilateral. In this situation there is no feedback from the output port and the input and output reflection coefficients are, respectively, independent of the load and source reflection coefficients.

Substitution into (3.3.1) and (3.3.2) for $S_{12} = 0$, gives the simple relations

$$\Gamma_{INU} = S_{11}$$

for the input reflection coefficient and

$$\Gamma_{OTU} = S_{22}$$

for the output reflection coefficient.

For the input stability independent of Z_S, the condition $\left|\Gamma_{IN}\right| < 1$, reduces to

$$\left|S_{11}\right| < 1. \qquad (4.4.1)$$

Similarly, the condition $\left|\Gamma_{OT}\right| < 1$ for the output stability independent of Z_L, reduces to

$$\left|S_{22}\right| < 1. \qquad (4.4.2)$$

For a unilateral network the stability of the input circuit is independent of Z_L, and the stability of the output circuit is independent of Z_S. Hence, if the input circuit is stable for all Z_S, it is also stable for all Z_L. Hence conditions (4.4.1) and (4.4.2) are the necessary and sufficient conditions for unilateral network to be unconditionally stable.

This can also be verified by considering the limiting case of the unconditional stability conditions ($K > 1$ and $\left|\Delta_S\right| < 1$) for bilateral networks as $S_{12} \to 0$.

With $S_{12} \to 0$, the above relations can be written respectively as

$$(1 - \left|S_{11}\right|^2)(1 - \left|S_{22}\right|^2) > 0 \qquad (4.4.3)$$

and

$$\left|S_{11}S_{22}\right| < 1. \qquad (4.4.4)$$

Clearly when conditions (4.4.1) and (4.4.2) are satisfied, both (4.4.3) and (4.4.4) are also satisfied. Conversely if (4.4.4) is satisfied, either or both $1 - \left|S_{11}\right|^2$ or $1 - \left|S_{22}\right|^2$ have to be greater than zero. If (4.4.3) is also satisfied, both expressions should have the same sign and hence both $1 - \left|S_{11}\right|^2$ and $1 - \left|S_{22}\right|^2$ are greater than zero.

As we have seen before, for the case of unilateral networks, the input and output circuits can be unstable independently. Hence, for a 2-port network to be unconditionally stable, the two conditions $|S_{11}| < 1$ and $|S_{22}| < 1$, should simultaneously hold. In cases when $|S_{11}| > 1$, it is still possible to have stability for certain values of Z_S, and similarly for the case $|S_{22}| > 1$, stability is possible for certain values of Z_L.

To find the values of Z_S that could lead to a stable condition in the cases of potentially unstable networks, we consider the fundamental stability condition

$$R_S + R_{IN} > 0$$

where $R_S = \text{Re}(Z_S)$ and

$$R_{IN} = \text{Re}(Z_{IN}) = \frac{1}{2}(Z_{IN} + Z_{IN}^*) = \frac{1}{2}\left[\frac{1+S_{11}}{1-S_{11}} + \frac{1+S_{11}^*}{1-S_{11}^*}\right]R_0 = \frac{1-|S_{11}|^2}{|1-S_{11}|^2}R_0.$$

In the above equation for R_{IN} the denominator is always positive, and for the case of potentially unstable unilateral network, the numerator is negative. Hence, the input resistance R_{IN} is negative. For stability therefore, R_S should be greater than the magnitude of R_{IN}.

In the Γ_S-plane, if we locate the circle $r_S = |r_{IN}| = |R_{IN}| / R_0$, the values of source impedances within this circle have a resistive part greater than $|r_{IN}|$ and hence create a stable situation (shaded area of Fig. (4.4.1)). Any normalized source impedance outside this circle, however, has normalized resistance less than $|r_{IN}|$ and is unstable. The unshaded area of Fig. (4.4.1) represents this unstable region.

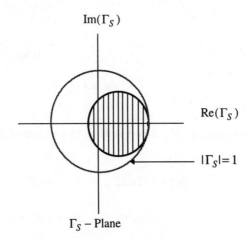

Γ_S – Plane

Fig. (4.4.1) Shaded area is stable

To find r_{IN}, however, we have to locate $\Gamma_{IN} = S_{11}$ in the complex Γ_{IN}-plane. As $|S_{11}| > 1$, this point will be outside the circle $|\Gamma_{IN}| = 1$, as shown in Fig. (4.4.2a). For convenience, therefore, we locate $1/S_{11}^{*}$ in the Δ_{IN}-plane ($\Delta_{IN} = 1/\Gamma_{IN}^{*}$), as discussed in Section (3.2) and shown in Fig. (4.4.2b). Once r_{IN} is located in Δ_{IN}-plane, we can locate an identically situated point $r_S = |r_{IN}|$ in the Γ_S-plane and find the stable and unstable regions.

We should emphasize that the stable region of Fig. (4.4.1) is different from the stable regions defined in Section (4.2) as shown in Fig (4.2.1). The latter refer to the values of Γ_L that lead to input stability for *all* values of Z_S.

The stability of the output circuit can be similarly considered by drawing an appropriate circle in Γ_L-plane, the stable and unstable regions can then be drawn. Although the input and output of unilateral 2-port networks can be considered separately, a network is stable if both ports are stable.

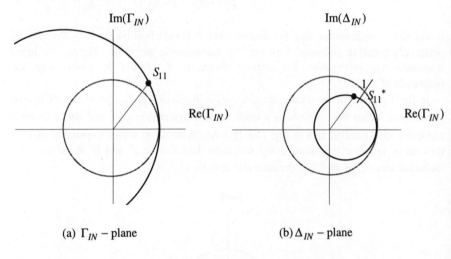

(a) Γ_{IN} – plane   (b) Δ_{IN} – plane

Fig. (4.4.2)

Example (4.4.1)

The scattering parameters of a 2-port network in a system of impedances $Z_1 = Z_2 = 50$ are given as

$$S_{11} = 0.72\angle 32°, \qquad S_{12} = 0\angle 0°,$$

$$S_{21} = 3.00\angle 0°, \qquad S_{22} = 1.46\angle 65°.$$

Discuss the stability of the unilateral 2-port network.

Solution

We consider the input port stability and the output port stability separately.

With $|S_{11}| < 1$, the input network is stable for all passive source impedances. The input impedance is given by

$$Z_{IN} = \frac{1 + S_{11}}{1 - S_{11}} R_{01} = 81 + j128$$

and has a positive resistive part. This ensures that for any passive source impedance, the net resistance of input circuit is positive and the input port is stable.

For the output port, however, $|\Gamma_{OT}| = |S_{22}| > 1$. In this case

$$Z_{OT} = \frac{1 + S_{22}}{1 - S_{22}} R_0 = -30 + j70$$

with a negative resistive part. Hence if the net resistance of the output port is to be greater than zero, the resistive part of the load impedance should exceed +30.

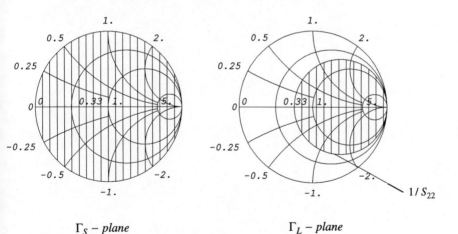

$\Gamma_S - plane$ $\Gamma_L - plane$

To find the boundary of the stable and unstable regions, we draw the circle of constant normalized resistance $r_L = 30/50 = 0.6$ in the Γ_L-plane, as shown. This circle passes through a corresponding point $\Delta_{OT} = 1/S_{22}^* = 0.685\angle 65°$ in Δ_{OT}-plane. Inside this circle, r_L is greater than 0.6 and the output port is stable. Outside the circle, the net resistance of the output port is negative and the output port is unstable.

For a load impedance of $Z_L = 30 - j70$, $Z_L + Z_{OT} = 0$ and the output circuit oscillates. In this situation $z_L = .6 - j1.4 (\Gamma_L = 0.685\angle -65° = 1/S_{22})$.

Chapter 5

Power Considerations of a 2-port Network

In this chapter we define and derive a set of expressions for the powers associated with a 2-port network. These include the available power from the source, power delivered to the load by the source and network, the maximum available power and the net power dissipated or generated within the network itself. It is assumed that the network is connected to a source with passive impedance at its input port and to an either passive or active circuit at its output port. We also define the network power gains and their dependence on the scattering parameters and the input and output impedances. Circles of constant operating power gain for bilateral networks and circles of constant transducer power gain for unilateral networks are given and discussed. Finally we consider the power relations in passive, lossless and reciprocal networks.

5.1 Definitions and Expressions for Power in a 2-port Network

Incident, reflected and delivered powers

Consider a 2-port network connected at port 1 to a source of emf E_1 and impedance Z_{S1}, and at port 2 to a source of emf E_2 and impedance Z_{S2}, as shown in Fig. (5.1.1). Current through and voltage across port 1 is denoted by I_1 and V_1, and for port 2 by I_2 and V_2. Taking the reference impedances as R_0 the following general power terms can be defined.

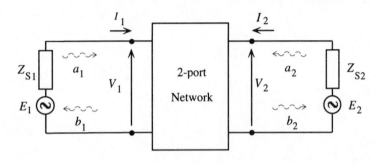

Fig. (5.1.1)

We define $a_1 a_1^*$ as the *incident power* on port 1 and $a_2 a_2^*$ as the *incident power* on port 2. Similarly we define $b_1 b_1^*$ as the *reflected power* from port 1 and $b_2 b_2^*$ as the *reflected power* from port 2. The difference between the incident power on port 1 and the reflected power from the same port

$$a_1 a_1^* - b_1 b_1^* \tag{5.1.1}$$

is the power delivered to port 1 of the network. Similarly the difference

$$a_2 a_2^* - b_2 b_2^* \tag{5.1.2}$$

is the power delivered to port 2 of the network.

Although (5.1.1) and (5.1.2) always hold, they can be easily verified for the case of $Z_{S1} = Z_{S2} = R_0$. Substituting for the wave amplitudes in terms of total voltages and total currents as given by (2.1.15) and (2.1.16) respectively, we can write (5.1.1) as

$$a_1 a_1^* - b_1 b_1^* = [(V_1 + R_0 I_1)(V_1^* + R_0 I_1^*) - (V_1 - R_0 I_1)(V_1^* - R_0 I_1^*)] / 4R_0$$
$$= (V_1 I_1^* + V_1^* I_1) / 2 \tag{5.1.3}$$

which is the power delivered to port 1. Equation (5.1.2) can be verified similarly.

Power definitions for a network connected to a generator at one port

We now apply the general power expressions to the situation where we have a single generator of emf E_S and impedance Z_S connected to port 1 or the input port of a 2-port network. Port 2 or the output port is terminated by an impedance Z_L. The circuit is shown in Fig. (5.1.2).

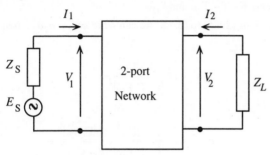

Fig. (5.1.2)

For this particular situation we can define the following power expressions:

(1) Power delivered to the input port by the source. This is termed the input power (P_{IN}).

(2) Power delivered to the input port by the source, when the input impedance is a conjugate of the source impedance. This is the available power from the source (P_{AVS}).

(3) Power delivered to the load by the source and network (P_L).

(4) Power delivered to the load by the source and network when the input impedance is a conjugate match of the source impedance ($P_{\Gamma_{IN} \text{ conj. } \Gamma_S}$).

(5) Power delivered to the load by the source and network when the output impedance is a conjugate of the load impedance. This is the power available from the source and network (P_{AVN}).

(6) Power delivered to the load by the source and the network for simultaneous conjugate matching conditions. This is the maximum available power (P_{MAX}).

(7) The net power delivered to the 2-port network (P_{NWK}).

We should note, however, that P_{MAX} is only defined when the network is unconditionally stable. Power for a potentially unstable network may reach very high values when the load termination is close to the terminations producing potential instability.

Expressions for power

The expressions for power are given here in terms of the scattering parameters and the reflection coefficients $\Gamma_L, \Gamma_S, \Gamma_{IN}$ and Γ_{OT}.

From the expressions (2.4.5) to (2.4.8) we can write $a_1 a_1^*$, $a_2 a_2^*$, $b_1 b_1^*$ and $b_2 b_2^*$ as

$$a_1 a_1^* = e_S^2 \frac{1 - |\Gamma_S|^2}{|1 - \Gamma_S \Gamma_{IN}|^2}, \tag{5.1.4}$$

$$a_2 a_2^* = e_S^2 \frac{1 - |\Gamma_S|^2}{|1 - \Gamma_S \Gamma_{IN}|^2} \frac{|\Gamma_L S_{21}|^2}{|1 - \Gamma_L S_{22}|^2}, \tag{5.1.5}$$

$$b_1 b_1^* = e_S^2 \frac{1 - |\Gamma_S|^2}{|1 - \Gamma_S \Gamma_{IN}|^2} |\Gamma_{IN}|^2, \tag{5.1.6}$$

$$b_2 b_2^* = e_S^2 \frac{1 - |\Gamma_S|^2}{|1 - \Gamma_S \Gamma_{IN}|^2} \frac{|S_{21}|^2}{|1 - \Gamma_L S_{22}|^2} \tag{5.1.7}$$

where

$$e_S = \frac{E_S}{2\sqrt{R_S}} = \frac{b_S}{\sqrt{1 - |\Gamma_S|^2}}.$$

From the expressions (5.1.4) to (5.1.7) and the above definitions, the following expressions for power can be found.

(1) Input power

This is defined as the power delivered to the input port and is given by

$$P_{IN} = a_1 a_1^* - b_1 b_1^*.$$

Substituting for $a_1 a_1^*$ and $b_1 b_1^*$ as given by (5.1.4) and (5.1.6) we have

$$P_{IN} = e_S^2 \left[\frac{(1 - |\Gamma_S|^2)(1 - |\Gamma_{IN}|^2)}{|1 - \Gamma_S \Gamma_{IN}|^2} \right]. \tag{5.1.8}$$

(2) Power available from the source

Power available from the source is the power delivered to the input port, when the input port is a conjugate match of the source. This can be found by substituting Γ_S^* for Γ_{IN} in (5.1.8). Hence the power available from the source is given as

$$P_{AvS} = e_S^2. \tag{5.1.9}$$

(3) Power dissipated in the load

Power dissipated in the load is the difference between power incident on and reflected from the load. This is given as

$$P_L = b_2 b_2^* - a_2 a_2^*$$

and after substitution, we have

$$P_L = e_s^2 \frac{1 - |\Gamma_S|^2}{|1 - \Gamma_S \Gamma_{IN}|^2} |S_{21}|^2 \frac{1 - |\Gamma_L|^2}{|1 - \Gamma_L S_{22}|^2}. \tag{5.1.10a}$$

Substituting for Γ_{IN} and Γ_{OT} from (3.3.1a) and (3.3.2a), respectively, it can be shown that

$$|1 - \Gamma_S \Gamma_{IN}|^2 |1 - \Gamma_L S_{22}|^2 = |1 - \Gamma_S S_{11}|^2 |1 - \Gamma_L \Gamma_{OT}|^2.$$

Hence (5.1.10a) can be written alternatively as

$$P_L = e_s^2 \frac{1 - |\Gamma_S|^2}{|1 - \Gamma_S S_{11}|^2} |S_{21}|^2 \frac{1 - |\Gamma_L|^2}{|1 - \Gamma_L \Gamma_{OT}|^2} \tag{5.1.10b}$$

(4) Power dissipated in the load with the input impedance a conjugate match of the source impedance

When the input impedance is a conjugate match of the source impedance, the expression for the power delivered to the load can be found by substituting Γ_{IN}^{*} for Γ_S in (5.1.10a) and hence this power is given as

$$P_{\Gamma_{IN} \text{ conj. } \Gamma_S} = e_S^2 \frac{1}{1 - |\Gamma_{IN}|^2} |S_{21}|^2 \frac{1 - |\Gamma_L|^2}{|1 - \Gamma_L S_{22}|^2}. \tag{5.1.11}$$

In this case $e_S = E_S / (2\sqrt{R_{IN}})$.

(5) Power available from the source and network

When the load impedance is a conjugate of the output impedance, power available from the source and the network P_{AvSN} is found by substituting Γ_L^{*} for Γ_{OT} in (5.1.10b) and hence we have

$$P_{AvSN} = e_S^2 \frac{1 - |\Gamma_S|^2}{|1 - \Gamma_S S_{11}|^2} |S_{21}|^2 \frac{1}{1 - |\Gamma_{OT}|^2}. \tag{5.1.12}$$

(6) Maximum available power

Power P_{MAX} delivered to the load when both source and load impedances are simultaneous conjugates of the input and output impedances, can be found by substituting $\Gamma_{IN} = \Gamma_{SMAX}^{*}$ and $\Gamma_L = \Gamma_{LMAX}$ in (5.1.11) or $\Gamma_{OT} = \Gamma_{LMAX}^{*}$ and $\Gamma_S = \Gamma_{SMAX}$ in (5.1.12). The two relevant expressions are

$$P_{MAX} = e_S^2 \frac{1}{1 - |\Gamma_{SMAX}|^2} |S_{21}|^2 \frac{1 - |\Gamma_{LMAX}|^2}{|1 - \Gamma_{LMAX} S_{22}|^2} \tag{5.1.13a}$$

and

$$P_{MAX} = e_S^2 \frac{1 - |\Gamma_{SMAX}|^2}{|1 - \Gamma_{SMAX} S_{11}|^2} |S_{21}|^2 \frac{1}{1 - |\Gamma_{LMAX}|^2}. \tag{5.1.13b}$$

Substituting in (5.1.13a) for Γ_{SMAX} in terms of Γ_{LMAX} or in (5.1.13b) for Γ_{LMAX} in terms of Γ_{SMAX} as given by (3.4.3) and (3.4.4), we have two alternative expressions for the maximum available power in the forms of

$$P_{MAX} = e_S^2 \frac{1 - |\Gamma_{LMAX}|^2}{\left|1 - \Gamma_{LMAX}\, S_{22}\right|^2 - \left|S_{11} - \Gamma_{LMAX}\, \Delta_S\right|^2} |S_{21}|^2 \qquad (5.1.14a)$$

and

$$P_{MAX} = e_S^2 \frac{1 - |\Gamma_{SMAX}|^2}{\left|1 - \Gamma_{SMAX} S_{11}\right|^2 - \left|S_{22} - \Gamma_{SMAX}\Delta_S\right|^2} |S_{21}|^2. \qquad (5.1.14b)$$

Substitution for Γ_{LMAX} as given by equation (3.4.14a) in (5.1.14a) or Γ_{SMAX} as given by (3.4.15a) in (5.1.14b) we finally have

$$P_{MAX} = e_S^2 \frac{|S_{21}|}{|S_{12}|}\left[K - \sqrt{(K^2 - 1)}\right] \qquad (5.1.15)$$

as the maximum available power when both input and output are simultaneously matched. For conjugate matching to be realized we require $K > 1$ and for maximum power to be delivered to the load, we require the unconditional stability or the additional condition of $|\Delta_S| < 1$.

If $K > 1$ and $|\Delta_S| > 1$, the conjugate matching is possible, but the network is potentially unstable. This situation, however, leads to a minimum and not a maximum power delivered to the load, as we shall see in section (5.3). The power delivered to the load is given by

$$P_{MIN} = e_S^2 \frac{|S_{21}|}{|S_{12}|}\left[K + \sqrt{(K^2 - 1)}\right]. \qquad (5.1.16)$$

In (5.1.15) $e_S = E_S / (2\sqrt{R_{SMAX}})$ and in (5.1.16) $e_S = E_S / (2\sqrt{R_{SMIN}})$ with $R_{SMAX} = \text{Re}(Z_{SMAX})$ and $R_{SMIN} = \text{Re}(Z_{SMIN})$.

(7) Power delivered to the network

The total power delivered to a 2-port network P_{NWK} is given by

$$P_{NWK} = P_{IN} - P_L.$$

Substituting for P_L and P_{IN} from Equations (5.1.8) and (5.1.10a) respectively, we have

$$P_{NWK} = e_S^2 \frac{1-|\Gamma_S|^2}{|1-\Gamma_S\Gamma_{IN}|^2}\left[1-|\Gamma_{IN}|^2 - |S_{21}|^2 \frac{1-|\Gamma_L|^2}{|1-\Gamma_L S_{22}|^2}\right]. \qquad (5.1.17)$$

For the particular situation of the input and output impedances equal to the resistive reference R_0, $\Gamma_S = \Gamma_L = 0$, and

$$P_{NWR} = e_S^2\left[1-|S_{11}|^2 - |S_{21}|^2\right] \qquad (5.1.18)$$

where $e_S^2 = E_S^2 / \left(2\sqrt{R_0}\right)$.

Example (5.1.1)

The scattering parameters of a 2-port network is measured in a system of reference impedances $R_0 = 50$ and given as

$$S_{11} = 0.40 + j0.20, \qquad S_{12} = 0.10 + j0.05,$$

$$S_{21} = 3.00 - j0.50, \qquad S_{22} = 0.30 - j0.10.$$

The network is terminated at its input port by a generator of emf 5 and internal impedance $Z_S = 30 + j20$ and at its output port by a load of $Z_L = 70 - j\,30$. Calculate all the defined power terms. For the case of load impedance conjugate matching, replace the load by its appropriate value. For the case of the source impedance a conjugate match of the input impedance or the case of simultaneous conjugate matching, assume a lossless circuit transforming the source impedance of the generator to the required impedance, with total available power from the source unaffected ($e_S{}^2$ assumed unchanged).

Solution

(1) Input power

In this case $Z_S = 30 + j20 (\Gamma_S = -0.176 + j0.294)$, $Z_L = 70 - j\,30 (\Gamma_L = 0.216 - j0.196)$ and $Z_{IN} = 130.8 + j52.5 (\Gamma_{IN} = 0.490 + j0.148)$. Using (5.1.8) and with $e_S^2 = E_S/(4R_S) = 0.208$, $1-|\Gamma_S|^2 = 0.883$, $1-|\Gamma_{IN}|^2 = 0.738$ and $|1-\Gamma_S\Gamma_{IN}| = 1.136$, the input power can easily be calculated as $P_{IN} = 0.105$.

(2) Power available from the source

In this case $Z_S = 30 + j20$ and $R_S = 30$. The available power from the source is given by (5.1.9) as $P_{AvS} = e_S^2 = E_S/(4R_S) = 0.208$.

(3) Power dissipated in the load

In this case $Z_S = 30 + j20(\Gamma_S = -0.176 + j0.294)$ and $Z_L = 70 - j\,30$ ($\Gamma_L = 0.216 - j0.196$). The power dissipated in the load is given by (5.1.10b). With $|S_{21}|^2 = 9.25$, $1-|\Gamma_S|^2 = 0.883$, $1-|\Gamma_L|^2 = 0.915$, $|1-\Gamma_S S_{11}| = 1.13$, $|1-\Gamma_L \Gamma_{OT}| = 0.961$, and $e_S^2 = 0.208$ we have $P_L = 1.31$.

(4) Power dissipated in the load with input impedance conjugate match to the source impedance

With a load of $Z_L = 70 - j\,30(\Gamma_L = 0.216 - j0.196)$, the input impedance is given as $Z_{IN} = 130.8 + j52.5(\Gamma_{IN} = 0.490 + j0.148)$. For the source to be a conjugate match of the input impedance we need $Z_S = 130.8 - j52.5$ ($\Gamma_S = 0.490 - j0.148$). Using (5.1.11) and with $|S_{21}|^2 = 9.25$, $1-|\Gamma_L|^2 = 0.915$, $|1-\Gamma_L S_{22}| = 0.958$, $1-|\Gamma_{IN}|^2 = 0.738$ and $e_S^2 = 0.208$, we find the required power to be equal to 2.60.

(5) Power available from the source and network

In this case $Z_S = 30 + j20$ ($\Gamma_S = -0.176 + j0.294$), $Z_{OT} = 77.8 - j6.05$ ($\Gamma_{OT} = 0.219 - j0.037$), $Z_L = 77.8 + j6.05$ ($\Gamma_L = 0.219 + j0.037$). Using (5.1.12) and with $1-|\Gamma_S|^2 = 0.883$, $|1-\Gamma_S S_{11}| = 1.13$, $1-|\Gamma_{OT}|^2 = 0.951$, $|S_{21}|^2 = 9.25$ and $e_S^2 = 0.208$, we find $P_{AvSN} = 1.40$.

(6) Maximum available power

As $K = 1.089 > 1$ and $|\Delta_S| = 0.202 < 1$, conjugate matching for maximum power delivered to the load is possible. For the simultaneous conjugate matching $Z_{SMAX} = 89.2 - j147.5$ ($\Gamma_{SMAX} = 0.662 - j0.359$), $Z_{LMAX} = 184.8 + j132.3$ ($\Gamma_{LMAX} = 0.677 + j0.182$). With $e_S^2 = 0.208$, $|S_{21}|/|S_{12}| = 9.62$, equation (5.1.15) gives the maximum available power as $P_{MAX} = 3.72$.

(7) Power delivered to the network

Using (5.1.17) and with $1-|\Gamma_S|^2 = 0.883$, $1-|\Gamma_{IN}|^2 = 0.738$, $1-|\Gamma_L|^2 = 0.915$, $|1-\Gamma_S \Gamma_{IN}| = 1.136$, $|1-\Gamma_L S_{22}| = 0.958$, $|S_{21}|^2 = 9.25$ and $e_S^2 = 0.208$, we have $P_{NWK} = P_{IN} - P_L = -1.21$. Hence a net power is generated inside the network.

5.2 Definition and Expressions for Power Gain

A ratio of any of the two power expressions given in the previous section is a dimensionless parameter and can be defined as a power ratio or a power gain. Of these, however, only four have useful interpretations and we confine ourselves to these four parameters. We use the term power gain for both cases of power ratios greater or less than unity.

The four power gains definitions are given below:

(1) The transducer power gain G_T, as the ratio of the power delivered to the load P_L to that of the power available from the source P_{AVS}.

(2) Operating power gain G_P as the ratio of the power delivered to the load P_L to the power delivered to the input port P_{IN}.

(3) Available power gain G_{Av} as the ratio of the power available from the source and network P_{AvSN}, to that of the power available from the source P_{AvS}.

(4) Maximum available power gain G_{MAX}, as the ratio of the power delivered to the load under simultaneous matched conditions P_{MAX}, to that of the available power from the source P_{AvS} for unconditionally stable networks.

Expressions for power gains

The expressions for power gains can easily be written as

(1) Transducer power gains

From (5.1.9) and (5.1.10a) we have

$$G_T = \frac{P_L}{P_{AvS}} = \frac{1-|\Gamma_S|^2}{|1-\Gamma_{IN}\Gamma_S|^2}|S_{21}|^2\frac{1-|\Gamma_L|^2}{|1-\Gamma_L S_{22}|^2} \qquad (5.2.1a)$$

or from (5.1.9) and (5.1.10b)

$$G_T = \frac{P_L}{P_{AvS}} = \frac{1-|\Gamma_S|^2}{|1-\Gamma_S S_{11}|^2}|S_{21}|^2\frac{1-|\Gamma_L|^2}{|1-\Gamma_{OT}\Gamma_L|^2}. \qquad (5.2.1b)$$

(2) Operating power gain

From (5.1.10a) and (5.1.8)

$$G_P = \frac{P_L}{P_{IN}} = \frac{1}{1-|\Gamma_{IN}|^2}|S_{21}|^2 \frac{1-|\Gamma_L|^2}{|1-\Gamma_L S_{22}|^2}. \qquad (5.2.2)$$

(3) Available power gain

From (5.1.12) and (5.1.9)

$$G_{Av} = \frac{P_{AvSN}}{P_{AvS}} = \frac{1-|\Gamma_S|^2}{|1-\Gamma_S S_{11}|^2}|S_{21}|^2 \frac{1}{1-|\Gamma_{OT}|^2}. \qquad (5.2.3)$$

(4) Maximum available power gain

From (5.1.13b) and (5.1.9)

$$G_{MAX} = \frac{P_{MAX}}{P_{AvS}} = \frac{1-|\Gamma_{SMAX}|^2}{|1-\Gamma_{SMAX} S_{11}|^2}|S_{21}|^2 \frac{1}{1-|\Gamma_{LMAX}|^2}. \qquad (5.2.4a)$$

Substituting for Γ_{SMAX} and Γ_{LMAX}, we find

$$G_{MAX} = \frac{|S_{21}|}{|S_{12}|}\left[K - \sqrt{(K^2-1)}\right] \qquad (5.2.4b)$$

which is only applicable to unconditionally stable networks with $K > 1$ and $|\Delta_S| < 1$.

Example (5.2.1)

For Example (5.1.1) of the previous section, calculate all the defined terms for power gain.

(1) The transducer power gain

$$G_T = \frac{P_L}{P_{AvS}} = \frac{1.31}{0.208} = 6.29 \quad \text{or} \quad 8 \quad \text{dB}$$

(2) Operating power gain

$$G_P = \frac{P_L}{P_{IN}} = \frac{1.31}{0.105} = 12.5 \quad \text{or} \quad 11 \quad \text{dB}$$

(3) Available power gain

$$G_{Av} = \frac{P_{AvN}}{P_{AvS}} = \frac{1.40}{0.208} = 6.73 \quad \text{or} \quad 8.3 \quad \text{dB}$$

(4) Maximum available power gain

$$G_{MAX} = \frac{P_{MAX}}{P_{AvS}} = \frac{3.72}{0.208} = 17.9 \quad \text{or} \quad 12.5 \quad \text{dB}$$

5.3 Circles of Constant Operating Power Gain for Bilateral Networks

In the previous section we have derived the required expressions for the transducer and operating power gains of a 2-port network. These were given by (5.2.1) and (5.2.2) in terms of the scattering parameters of the network. A plot in a complex Γ_S or Γ_L plane of the locus of the points of constant transducer power gain or operating power gain, gives a desirable representation of the effect of the source or load terminations on the network power gain.

It is clear from these equations, however, that the transducer power gain is dependent on both the source and load impedances, while the operating power gain is only a function of Γ_L and not Γ_S. The locus of the points of constant transducer power gain in the Γ_L-plane can only be drawn for each value of Γ_S, and hence we may have an infinite number of such set of circles.

For the operating power gain being independent of Γ_S, however, only a single set of curves in Γ_L-plane is required. We examine this set of curves in this section. It is also assumed that the network is bilateral where both $|S_{12}| \neq 0$ and $|S_{21}| \neq 0$. The case of $|S_{12}|$ being negligibly small ($|S_{21}| \neq 0$) will be considered in the next section, where we examine the transducer power gain of unilateral networks.

To find the locus of the points of constant operating gain, we first normalize the operating power gain as

$$g_P = \frac{|S_{12}|}{|S_{21}|} G_P \tag{5.3.1}$$

or from (5.2.2)

$$g_P = \frac{1}{1 - |\Gamma_{IN}|^2} |S_{12} S_{21}| \frac{1 - |\Gamma_L|^2}{|1 - \Gamma_L S_{22}|^2} .$$

With substitution for Γ_{IN} as

$$\Gamma_{IN} = \frac{S_{11} - \Gamma_L \Delta_S}{1 - \Gamma_L S_{22}},$$

we can write g_p as

$$g_p = \frac{|S_{12} S_{21}|(1 - |\Gamma_L|^2)}{|1 - \Gamma_L S_{22}|^2 - |S_{11} - \Gamma_L \Delta_S|^2}. \qquad (5.3.2)$$

Expressing the denominator of the above equation as

$$|1 - \Gamma_L S_{22}|^2 - |S_{11} - \Gamma_L \Delta_S|^2$$

$$= 1 - |S_{11}|^2 - (S_{22} - \Delta_S S_{11}^*)\Gamma_L - (S_{22}^* - \Delta_S S_{11}^*)\Gamma_L^*$$

$$+ (|S_{22}|^2 - |\Delta_S|^2)\Gamma_L \Gamma_L^*,$$

(5.3.2) can be written in the expanded form

$$\Gamma_L \Gamma_L^* - \frac{C_L^* g_p}{1 + (B_L - K)g_p}\Gamma_L^* - \frac{C_L g_p}{1 + (B_L - K)g_p}\Gamma_L - \frac{1 - (B_L + K)g_p}{1 + (B_L - K)g_p} = 0. \qquad (5.3.3)$$

In (5.3.3) we have used the previously defined terms

$$K = \frac{1 - |S_{11}|^2 - |S_{22}|^2 + |\Delta_S|^2}{2|S_{12} S_{21}|},$$

$$B_L = \frac{1 - |S_{11}|^2 + |S_{22}|^2 - |\Delta_S|^2}{2|S_{12} S_{21}|},$$

$$C_L = \frac{S_{22} - \Delta_S S_{11}^*}{|S_{12} S_{21}|}$$

together with the following simple relations

$$B_L + K = \frac{1 - |S_{11}|^2}{|S_{12}S_{21}|},$$

$$B_L - K = \frac{|S_{22}|^2 - |\Delta_S|^2}{|S_{12}S_{21}|}.$$

The relation $K^2 - B_L^2 + C_L C_L^* = 1$ between K, B_L and C_L also holds.

As we have seen before, equations in the form of (5.3.3) can be written as

$$|\Gamma_L - \Gamma_{LCP}|^2 - \rho_{LCP}^2 = 0 \qquad (5.3.4)$$

and represent a set of circles in Γ_L-plane. The centers and radii of the circles are given respectively as

$$\Gamma_{LCP} = \frac{C_L^* g_p}{1 + (B_L - K)g_p} \qquad (5.3.5)$$

and

$$\rho_{LCP} = \frac{\sqrt{1 - 2Kg_p + g_p^2}}{\left|1 + (B_L - K)g_p\right|}. \qquad (5.3.6)$$

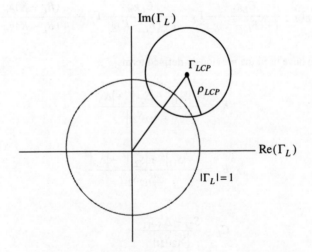

Fig. (5.3.1)

The centers of the constant gain circles lie on a line, passing through the origin and inclined by an angle

$$\phi_{LCP} = \tan^{-1} \frac{\text{Im}(C_L^*)}{\text{Re}(C_L^*)}.$$

The centers are at distances from the origin of the Γ_L-plane given as

$$|\Gamma_{LCP}| = \frac{\sqrt{1 + (B_L^2 - K^2)}\, g_p}{|1 + (B_L - K)g_p|}.$$

A typical circle is given in Fig. (5.3.1).

For further discussion, we find the intersection of these circles with the circle $|\Gamma_L| = 1$. Substituting $\Gamma_L = \Gamma_\Lambda = \exp(j\phi_\Lambda)$ and $\Gamma_{LCP} = |\Gamma_{LCP}|\exp(j\phi_p)$ in the equation for the circles of constant operating gain as given by (5.3.4), we have

$$1 - |\Gamma_{LCP}|\exp[j(\phi_P - \phi_\Lambda)] - |\Gamma_{LCP}|\exp[-j(\phi_P - \phi_\Lambda)] + |\Gamma_{LCP}|^2 - \rho_{LCP}^2 = 0$$

or hence

$$\cos(\phi_P - \phi_\Lambda) = \frac{1 + |\Gamma_{LCP}|^2 - \rho_{LCP}^2}{2|\Gamma_{LCP}|}. \tag{5.3.7}$$

From (5.3.7) it is clear that the intersection of the operating power gain circle and the circle $|\Gamma_L| = 1$ is impossible if $|\cos(\phi_P - \phi_\Lambda)| > 1$.

This condition can be written as

$$\frac{\left|1 + |\Gamma_{LCP}|^2 - \rho_{LCP}^2\right|}{2|\Gamma_{LCP}|} > 1 \tag{5.3.8}$$

which leads to situations similar to those of A1(a), A4(a), B1(a), B4(a), C1(a), C2(a) and C4(a) of Fig. (4.2.1a) and A1(b), A4(b), B1(b), B4(b), C1(b), C2(b) and C4(b) of Fig. (4.2.1b) of the stability circles.

Substituting for $|\Gamma_{LCP}|$ from (5.3.5) and ρ_{LCP} from (5.3.6) into (5.3.7), we finally have

$$\cos(\phi_P - \phi_\Lambda) = \frac{B_L}{\sqrt{[1 + (B_L^2 - K^2)]}} \tag{5.3.9}$$

with the term inside the square root sign always positive ($= C_L C_L^*$).

Hence the condition for no intersection is

$$\frac{B_L^2}{[1+(B_L^2 - K^2)]} > 1 \qquad (5.3.10)$$

which leads to the simple condition of $|K| > 1$ as in Chapter 4.

From equation (5.3.9) it is clear that for $|K| \le 1$, all circles of constant operating gain intersect the circle $|\Gamma_L| = 1$ at the same two points, independent of g_p. When $K = 1$, the two points coincide and all circles become tangential to the circle of $|\Gamma_L| = 1$.

It is of interest to consider the variation of Γ_{LCP} and ρ_{LCP}, with g_p. We first start with two extreme values of $g_p \to 0$ and $g_p \to \infty$.

For $g_p \to 0$ we have

$$\Gamma_{LCP} \to 0$$

$$\rho_{LCP} \to 1$$

or the circle of normalized operating power gain circle $g_p = 0$ coincides with the circle $|\Gamma_L| = 1$. It is obvious that loads that are purely reactive have Γ_L on the circle $|\Gamma_L| = 1$ and for these loads the operating power gain is zero.

For $g_p \to \infty$, we have

$$\Gamma_{LCP} \to \frac{C_L^*}{(B_L - K)} = \Gamma_{LCS} \quad (|\Gamma_{LCP}| \to \frac{\sqrt{[1 - (B_L^2 - K^2)]}}{|B_L - K|} = |\Gamma_{LCS}|),$$

$$\rho_{LCP} \to \frac{1}{|B_L - K|} = \rho_{LCS}$$

which are identical, respectively, to the center and radius of the Γ_L-plane stability circle given in Section (4.1). Hence the stability circles are the circles of infinite operating gain.

In Fig. (5.3.1) we plot $|\Gamma_{LCP}|$ and ρ_{LCP} as functions of g_p, with K as a parameter and for fixed values of B_L.

From (5.3.5) it is clear that

$$1 + (B_L^2 - K^2) > 0$$

or for a fixed value of B_L

$$|K| < \sqrt{(1 + B_L^2)}$$

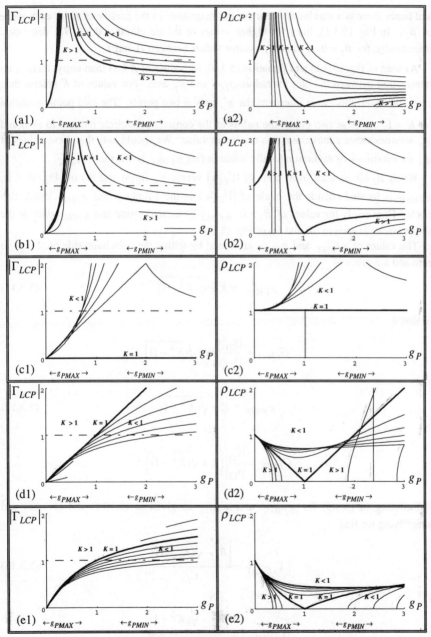

Fig. 5.3.1 (a)$B_L = -2$, (b)$B_L = -1$, (c)$B_L = 0$, (d)$B_L = 1$, (e)$B_L = 2$

and hence there is a maximum value for the magnitude of the parameter K for each value of B_L. In Fig. (5.3.1), lines for higher values of $|K|$ are eliminated as $|B_L|$ decreases. Specifically, for $B_L = 0$, K can only assume values between -1 and 1.

As seen in Fig. (5.3.1) and Equation (5.3.5), for values of K less than unity, there is no intersection between the plot for radius ρ_{LCP} and g_p axis. For values of K greater than unity, however, this curve intersects the g_p axis at two points. The two points coincide for $K = 1$. At these two points the radius of the constant gain circle reduces to zero and g_p assumes either a maximum or a minimum value. We should note that these values of g_p are maximum or minimum locally while in fact g_{PMAX} is less than g_{PMIN}.

When $B_L < 0$, as seen from plots of $|\Gamma_{LCP}|$ versus g_p in (a) and (b) of Fig. (5.3.1), g_{PMAX} is located outside the circle of $|\Gamma_L| = 1$ in the Γ_L-plane and g_{PMIN} inside this circle. Conversely for values of $B_L > 0$, g_{PMAX} is located inside and g_{PMIN} outside the above circle, as seen in (d) and (e) of Fig. (5.3.1).

The values of g_{PMAX} and g_{PMIN} are found by setting the right hand side of (5.3.6) to zero and are given as

$$g_{PMAX} = K - \sqrt{(K^2 - 1)} \tag{5.3.11}$$

or hence

$$G_{PMAX} = \frac{|S_{21}|}{|S_{12}|} \left[K - \sqrt{(K^2 - 1)} \right]$$

and

$$g_{PMIN} = K + \sqrt{(K^2 - 1)} \tag{5.3.12}$$

or

$$G_{PMIN} = \frac{|S_{21}|}{|S_{12}|} \left[K + \sqrt{(K^2 - 1)} \right].$$

Substituting in (5.3.5) for g_{PMAX} and g_{PMIN} as given in (5.3.11) and (5.3.12) and simplifying we find

$$\Gamma_{LCMAX} = \frac{\left[B_L - \sqrt{(K^2 - 1)} \right]}{C_L} \tag{5.3.13}$$

or

$$|\Gamma_{LCMAX}| = \frac{\left| B_L - \sqrt{(K^2 - 1)} \right|}{\sqrt{1 + (B_L^2 - K^2)}}$$

and similarly

$$\Gamma_{LCMIN} = \frac{\left[B_L + \sqrt{(K^2 - 1)} \right]}{C_L} \tag{5.3.14}$$

or

$$\left| \Gamma_{LCMIN} \right| = \frac{\left| B_L + \sqrt{(K^2 - 1)} \right|}{\sqrt{1 + (B_L^2 - K^2)}}.$$

Considering the points inside the circle $\left| \Gamma_L \right| = 1$, (5.3.13) applies when $B_L > 0$ and (5.3.14) applies when $B_L < 0$.

We can now see that the expressions for Γ_{LCMAX} and Γ_{LCMIN} as given by (5.3.13) and (5.3.14) are identical to the load reflection coefficients Γ_{LMAX} and Γ_{LMIN} as given by (3.4.14a) and (3.4.14b). Hence the load impedances for g_{PMAX} and g_{PMIN} are identical to the load impedances for the simultaneous conjugate matching. If the input port is appropriately terminated, the load reflection coefficient given by (5.3.13) for $B_L > 0$, produces a maximum transducer power gain for an unconditionally stable 2-port network. It is also clear that $G_{PMAX} = G_{MAX}$ and $G_{PMIN} = G_{MIN}$, as we have defined in the previous section.

Typical circles of constant operating power gain are given in the following example for similar situations to those shown in Fig. (4.2.1a) and Fig. (4.2.1b). These curves should also be compared with the plots of Fig. (5.3.1).

Example (5.3.1)

For the networks of Example (4.2.1), draw all possible circles of constant operating power gain with values of g_P given as

$$g_P = N / 5 \qquad \text{for} \quad 0 \le N < 5$$

$$g_P = 5 / (10 - N) \qquad \text{for} \quad 5 \le N \le 10$$

where N is an integer number. Indicate whether the conjugate matching is possible and whether the operating power gain, in case of conjugate matching, is a maximum or a minimum. This example should be considered in conjunction with Example (4.2.1), where we have found the centers and radii of the stability circles. In fact the stability circles are the circles of infinite operating gain or when $N = 10$ for the present example.

Solution

The circles of constant operating power gain are drawn in the Γ_L-plane for each of the above cases. For completeness the circles are drawn for both regions outside as well as inside the $|\Gamma_L| = 1$ circle. The regions of the Γ_L-plane that contain the circles of constant operating gain are stable, while the regions of the plane that are depleted of these circles are potentially unstable. These regions correspond to negative values of operating power gain, which implies a negative input power or a net power flow from the network to the source. In turn a negative input power implies an input impedance with negative real part, which can lead to instability of the input port.

It should also be noted that it may not be possible to draw the circles of constant operating power gain for all required values of N. As we have discussed and shown in Fig. (5.3.1), with $K > 1$, g_P assumes a local maximum g_{PMAX} and a local minimum g_{PMIN}. As noted before, $g_{PMIN} > g_{PMAX}$ and g_P cannot assume any value between these two limits.

For $N = 0$, $g_P = 0$ and the circle of constant operating power gain coincides with the circle of $|\Gamma_L| = 1$, denoted in the following plots as UC (unit circle). In this example for $N = 10$, g_P approaches infinity and the circle of constant operating power gain coincides with the stability circle. These circles are denoted by SC.

A1(a) $(K = 1.64, \quad B_L = 3.64, \quad |\Delta_S| = 0.25)$

In this case $K > 1$ and simultaneous conjugate matching can be realized with a load of $\Gamma_{LCMAX} = 0.688 \angle 60°$, giving a normalized gain of $g_{PMAX} = 0.34$. Simultaneous conjugate matching can also be realized for a minimum normalized operating power gain of $g_{PMIN} = 2.94$ with $\Gamma_{LCMIN} = 1.45 \angle 60°$. However, this value of the load reflection coefficient cannot be obtained with passive terminations. The circles of constant operating power gain can only be drawn for values of N given as $N = 0, 1, 9$ and $N = 10$.

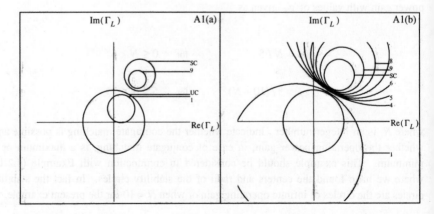

A1(b) ($K = -1.66$, $B_L = -3.68$, $|\Delta_S| = 0.85$)

In this case, as the network is potentially unstable for all passive values of the load impedance, no circle of constant operating gain is confined within the $|\Gamma_L| = 1$ circle. The circles of constant operating power gain for possible active loads with negative resistive impedance are as shown.

A2(a) ($K = 0.44$, $B_L = 1.45$, $|\Delta_S| = 0.42$)

With $|K| < 1$, all constant gain circles intersect the circle of $|\Gamma_L| = 1$ at two points. These two points, from Equation (5.3.9), have values of $\Gamma_L = 1 \angle 28°$ and $\Gamma_L = 1 \angle 92°$ in the Γ_L-plane. In this case the simultaneous conjugate matching is not possible.

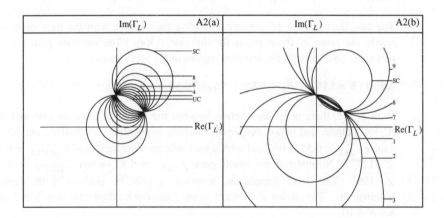

A2(b) ($K = -0.44$, $-B_L = 1.45$, $|\Delta_S| = 1.27$)

All constant gain circles intersect the circle of $|\Gamma_L| = 1$ at the same two points as A2(a). The regions of positive and negative operating power gains are, however, interchanged (negative gain circles not shown).

A3(a) ($K = 0.45$, $B_L = -0.07$, $|\Delta_S| = 0.69$)

With $|K| < 1$, there are two intersection points with the circle $|\Gamma_L| = 1$ at $\Gamma_L = 1 \angle -25°$ and $\Gamma_L = 1 \angle 145°$. Simultaneous conjugate matching is not possible.

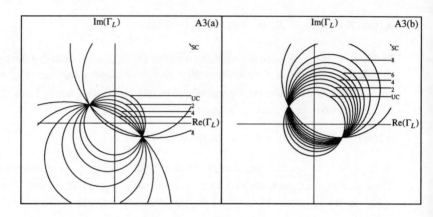

A3(b) ($K = -0.45$, $B_L = 0.07$, $\left|\Delta_S\right| = 1.16$)

The two intersections of constant operating gain circles with the unit circle are nearly the same as those given for the case A3(a). The constant gain circles, however, pass through the depleted regions of the previous case.

A4(a) ($K = 1.18$, $B_L = 0.85$, $\left|\Delta_S\right| = 0.50$)

With $K > 1$, there is no intersection between the circles of constant gain and the $\left|\Gamma_L\right| = 1$ circle, and hence, conjugate matching is possible. Normalized maximum gain $g_{PMAX} = 0.55$ is realized with a load reflection coefficient of $\Gamma_{LCMAX} = 0.38$ $\angle -120°$. Normalized minimum gain $g_{PMIN} = 1.82$, when $\Gamma_{LCMIN} = 2.62$ $\angle -120°$. This load impedance, however, cannot be realized with passive terminations. The circles of constant power gains can be drawn for $0 \le N \le 2$ and $8 \le N \le 10$.

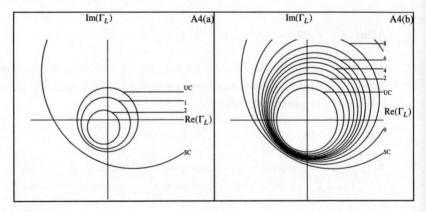

A4(b) ($K = -1.18$, $B_L = -0.85$, $|\Delta_S| = 0.49$)

In this case, as the network is potentially unstable for all passive load impedances, no circle of constant operating gain is confined within the $|\Gamma_L| = 1$ circle. The circles of constant gains shown, can only be realized with loads having negative resistive impedance or active loads.

B1(a) ($K = -1.22$, $B_L = 2.12$, $|\Delta_S| = 1.04$)

There is no intersection of the constant operating power gain circles with the $|\Gamma_L| = 1$ circle. The stability circle is inside $|\Gamma_L| = 1$ circle and all other constant gain circles lie between the two circles as shown. Simultaneous conjugate matching is not possible with passive terminations.

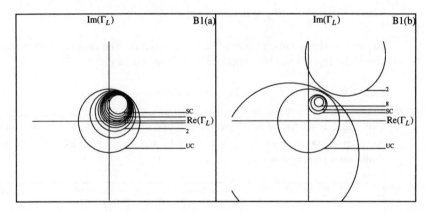

B1(b) ($K = 1.22$, $B_L = -2.12$, $|\Delta_S| = 1.43$)

As in case B1(a), there is no intersection between the circles of constant operating gain and the $|\Gamma_L| = 1$ circle. In this case however as $K > 1$, simultaneous conjugate matching is possible. For $g_{PMAX} = 0.52$, $\Gamma_{LCMAX} = 1.41 \angle 60°$ and is outside the $|\Gamma_L| = 1$ circle and hence cannot be realized with passive load impedances. However, for $g_{PMIN} = 1.91$, $\Gamma_{LCMIN} = 0.71 \angle 60°$, and is inside the $|\Gamma_L| = 1$ circle. The value of normalized gain increases from g_{PMIN} to infinity as we move from Γ_{LCMIN} in any direction towards the stability circle. It is only possible to draw circles of constant operating power gain for values of $0 \le N \le 2$ and $8 \le N \le 10$.

B2(a) ($K = -0.89$, $B_L = 1.11$, $|\Delta_S| = 0.92$)

With $|K| < 1$, all constant gain circles intersect the circle of $|\Gamma_L| = 1$ at two points. These points are $\Gamma_L = 1 \angle 37°$ and $\Gamma_L = 1 \angle 82°$.

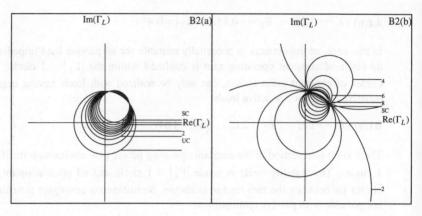

B2(b) ($K = 0.89$, $B_L = -1.11$, $|\Delta_S| = 1.34$)

The intersection points are nearly the same as B2(a). The constant gain circles are now in the regions that were depleted in the previous case.

B3(a) ($K = 0.81$, $B_L = -0.31$, $|\Delta_S| = 0.90$)

With $|K| < 1$, all the constant gain circles intersect the circle of $|\Gamma_L| = 1$ at two points given in this case at $\Gamma_L = 1 \angle -2.6°$ and $\Gamma_L = 1 \angle 123°$. Simultaneous conjugate matching is not possible.

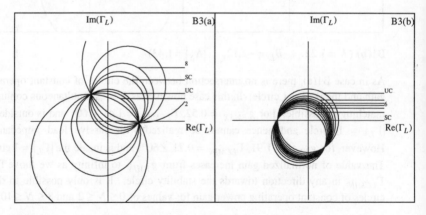

B3(b) ($K = -0.81$, $B_L = 0.31$, $|\Delta_S| = 1.40$)

The intersection points are roughly the same as B3(a). The constant gain circles are now in the regions that were depleted in the previous case.

B4(a) ($K = 1.12$, $B_L = 0.59$, $|\Delta_S| = 0.74$)

With $K > 1$, there is no intersection between the stability circle and the $|\Gamma_L| = 1$ circle. Conjugate matching is possible with a load reflection coefficient of Γ_{LCMAX} = 0.29 $\angle - 120°$ giving g_{PMAX} = 0.62. Also g_{PMIN} = 1.62 requires a load reflection coefficient of Γ_{LCMIN} = 3.44 $\angle - 120°$. The circles of constant operating power gain can be drawn for values of $0 \le N \le 3$ and $7 \le N \le 10$.

B4(b) ($K = -1.12$, $B_L = -0.60$, $|\Delta_S| = 0.75$)

The network is potentially unstable for all passive load impedances, hence no circle of constant operating gain is confined within the $|\Gamma_L| = 1$ circle. The circles of constant gain shown can only be realized with load impedances having negative resistive part.

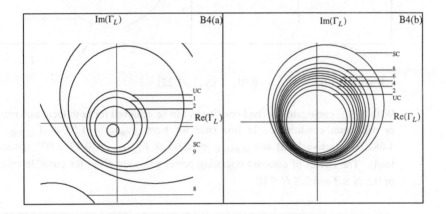

C1(a) ($K = -1.55$, $B_L = 1.78$, $|\Delta_S| = 1.40$)

With $|K| > 1$, and $|\Delta_S| > 1$ the stability circle is completely inside the $|\Gamma_L| = 1$ circle. the circles of constant operating gain lie in the region between the two circles as shown.

C1(b) ($K = 1.54$, $B_L = -1.78$, $|\Delta_S| = 1.78$)

With $K > 1$, when the network is conjugate matched with load reflection coefficients of Γ_{LCMAX} = 2.23 $\angle 60°$ and Γ_{LCMIN} = 0.45 $\angle 60°$, the normalized power gains will be given as $g_{PMAX} = 0.37$ and $g_{PMIN} = 2.72$ respectively. Hence g_{PMIN} is located inside the stability circle and g_{PMAX} outside the $|\Gamma_L| = 1$ circle. The circles of constant operating gain are given for $0 \leq N \leq 1$ and $9 \leq N \leq 10$.

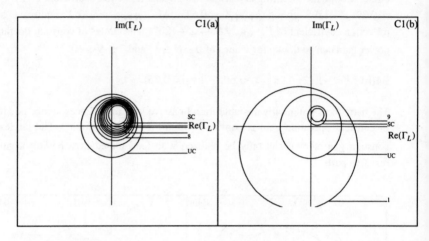

C2(a) ($K = 1.09$, $B_L = -0.91$, $|\Delta_S| = 1.25$)

With $K > 1$, conjugate matched conditions can be achieved for both local maximum or minimum conditions. In this case we have $g_{PMIN} = 1.53$ at $\Gamma_{LCMIN} = 1.68\angle60°$ (active load) and $g_{PMAX} = 0.66$ at $\Gamma_{LCMAX} = 0.59\angle60°$ (passive load). The circles of constant operating power gains are given for possible values of $0 \leq N \leq 3$ and $7 \leq N \leq 10$.

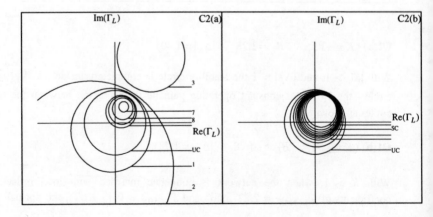

C2(b) ($K = -1.09$, $B_L = 0.91$, $|\Delta_S| = 1.15$)

In this case the inside of the stability circle is potentially unstable and all circles of constant power gain lie between the stability circle and the circle $|\Gamma_L| = 1$ as shown.

C3(a) ($K = 0.93$, $B_L = -0.32$, $|\Delta_S| = 0.92$)

With $|K| < 1$, all circles of constant operating power gain intersect the $|\Gamma_L| = 1$ circle at two points. These two points are $\Gamma_L = 1\angle 10°$ and $\Gamma_L = 1\angle 109°$. The circles of constant operating power gain are as shown.

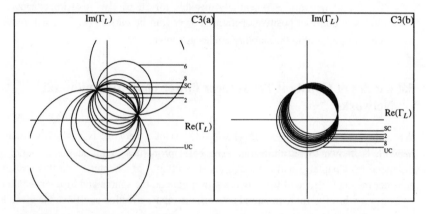

C3(b) ($K = -0.92$, $B_L = 0.32$, $|\Delta_S| = 1.35$)

The intersection points of all operating power gain circles are similar to the previous case. These circles, however, lie on the regions that were depleted in the previous case.

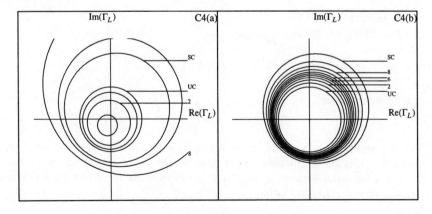

C4(a) ($K = 1.10$, $B_L = 0.51$, $|\Delta_S| = 0.73$)

The 2-port network is unconditionally stable and conjugate matching for maximum power gain is possible. g_{PMAX} = 0.65, which is realized with a load of reflection coefficient Γ_{LCMAX} = 0.25. g_{PMIN} = 1.55 at a load of   reflection coefficient Γ_{LCMIN} = 4.06 $\angle -120°$, which cannot be realized by any passive load. Circles of constant operating power gain are given for $0 \le N \le 3$ and $7 \le N \le 10$.

C4(b) ($K = -1.10$, $B_L = -0.51$, $|\Delta_S| = 0.71$)

The 2-port network is potentially unstable for all passive load impedances. All circles of constant positive operating power gain lie outside the $|\Gamma_L|$ = 1 circle and are confined within the stability circles as shown.

5.4 Circles of Constant Transducer Power Gain for Unilateral Networks

When the scattering parameter $|S_{12}| (|S_{12}|<|S_{21}|)$ of a network is negligibly small, the network is known to be unilateral. The basic properties of unilateral networks were discussed in Chapter 3 and their stability in Section 3 of Chapter 4. In this section we consider the dependence of the network power gain on the source and load terminations.

To review briefly, for a unilateral network the input reflection coefficient is given by

$$\Gamma_{INU} = S_{11}$$

and is independent of Γ_L. Similarly

$$\Gamma_{OTU} = S_{22}$$

and is independent of Γ_S.

The stability conditions are simply

$$|S_{11}| < 1$$

for the input port and

$$|S_{22}| < 1$$

for the output port. With the above stability conditions simultaneously satisfied, the network is unconditionally stable or it is stable for any source or load impedance.

For a unilateral network as $|S_{12}| \rightarrow 0$, the expressions for power gain as given in Section (5.2) reduce to the following:

(i) Transducer Power Gain

Substituting for $\Gamma_{INU} = S_{11}$ in (5.2.1a), or $\Gamma_{OTU} = S_{22}$ in (5.2.1b) we have

$$G_{TU} = \frac{P_L}{P_{AVS}} = \frac{1-|\Gamma_S|^2}{|1-S_{11}\Gamma_S|^2}|S_{21}|^2 \frac{1-|\Gamma_L|^2}{|1-S_{22}\Gamma_L|^2}. \tag{5.4.1}$$

In the above expressions U stands for the unilateral networks.

(ii) Operating Power Gain

Substituting for $\Gamma_{INU} = S_{11}$ in (5.2.2) we have

$$G_{PU} = \frac{P_L}{P_{IN}} = \frac{1}{1-|S_{11}|^2}|S_{21}|^2 \frac{1-|\Gamma_L|^2}{|1-S_{22}\Gamma_L|^2}. \tag{5.4.2}$$

When $|S_{11}|$ is greater than unity, the operating power gain becomes negative. In this case the input power is negative or more power is reflected back from the 2-port network than the power supplied by the source.

(iii) Available Power Gain

Substituting for $\Gamma_{OTU} = S_{22}$ in (5.2.3) we have

$$G_{AVU} = \frac{P_{AVN}}{P_{AVS}} = \frac{1-|\Gamma_S|^2}{|1-S_{11}\Gamma_S|^2}|S_{21}|^2 \frac{1}{1-|S_{22}|^2}. \tag{5.4.3}$$

The above equation is only applicable if $|S_{22}| < 1$.

(iv) Maximum Available Power Gain

Maximum available power is given by the expression

$$G_{MAXU} = \frac{P_{MAX}}{P_{AVS}} = \frac{1}{1-|S_{11}|^2}|S_{21}|^2 \frac{1}{1-|S_{22}|^2}. \tag{5.4.4}$$

The last equation is found by substituting $\Gamma_{SMAX} = S_{11}^*$ and $\Gamma_{LMAX} = S_{22}^*$ in (5.2.4a). Equation (5.2.4b) also reduces to (5.4.4) if we note that as $|S_{12}| \to 0$, $K \to \infty$, but the product $K \times |S_{12}|$ remains finite. To verify we write

$$G_{MAXU} = \frac{|S_{21}|}{|S_{12}|}\left[K - \left(K^2 - 1\right)^{1/2}\right]$$

$$= \frac{|S_{21}|}{|S_{12}|} K\left[1 - \left(1 - K^{-2}\right)^{1/2}\right]$$

$$= \frac{|S_{21}|}{|S_{12}|}\frac{1}{2K} \qquad K \to \infty$$

and finally substituting for K as

$$K = \frac{\left(1 - |S_{11}|^2\right)\left(1 - |S_{22}|^2\right)}{2|S_{12}S_{21}|},$$

we find the required expression. It should be noted that the maximum available power gain as given by the expression (5.4.4), is only applicable if both $|S_{11}| < 1$ and $|S_{22}| < 1$.

For the case of unilateral networks, the dependence of the transducer power gain on the reflection coefficients Γ_S and Γ_L can easily be investigated. This is because the expression for the transducer gain can be written as

$$G_{TU} = G_S(\Gamma_S)|S_{21}|^2 G_L(\Gamma_L) \tag{5.4.5}$$

where

$$G_S(\Gamma_S) = \frac{1 - |\Gamma_S|^2}{|1 - \Gamma_S S_{11}|^2} \tag{5.4.6}$$

and is independent of Γ_L and

$$G_L(\Gamma_L) = \frac{1 - |\Gamma_L|^2}{|1 - \Gamma_L S_{22}|^2} \tag{5.4.7}$$

and is independent of Γ_S.

The first and last terms of expression (5.4.5), can hence be investigated independently and as functions of Γ_S and Γ_L respectively. The middle term, however, is a constant of the network and cannot be affected by the changes in the source and load impedance.

As (5.4.6) and (5.4.7) have identical form we write

$$G_i(\Gamma_i) = \frac{(1-|\Gamma_i|^2)}{|1-\Gamma_i S_{ii}|^2} \quad i = 1,2 \tag{5.4.8}$$

where $G_i = G_S$, $\Gamma_i = \Gamma_S$ and $S_{ii} = S_{11}$ for $i = 1$ and $G_i = G_L$, $\Gamma_i = \Gamma_L$ and $S_{ii} = S_{22}$ for $i = 2$. To examine the effect of the source and load reflection coefficients on the gain factors $G_S(\Gamma_S)$ and $G_L(\Gamma_L)$, we plot the locus of the points of constant gain factor in Γ_S-plane and Γ_L-plane respectively. For $G_i(\Gamma_i)$ set to a constant in the form of

$$G_i(\Gamma_i) = \frac{g_i}{1-|S_{ii}|^2},$$

(5.4.8) can be written as

$$g_i(1-\Gamma_i S_{ii})(1-\Gamma_i^* S_{ii}^*) = (1-\Gamma_i \Gamma_i^*)(1-S_{ii}S_{ii}^*)$$

and rearranged as

$$\Gamma_i \Gamma_i^* - [\frac{g_i S_{ii}^*}{1-(1-g_i)S_{ii}S_{ii}^*}]\Gamma_i^* - [\frac{g_i S_{ii}}{1-(1-g_i)S_{ii}S_{ii}^*}]\Gamma_i = \frac{1-g_i - S_{ii}S_{ii}^*}{1-(1-g_i)S_{ii}S_{ii}^*}. \tag{5.4.9}$$

For each constant value of g_i, (5.4.9) is an equation of a circle in the complex Γ_i-plane. The center of the circle is at a position given by

$$\Gamma_{iC} = \frac{g_i S_{ii}^*}{1-(1-g_i)|S_{ii}|^2} \tag{5.4.10}$$

which lies on a line making an angle ϕ_{iC} with the $\text{Re}(\Gamma_i)$ axis given by

$$\phi_{iC} = \tan^{-1}\left[-\frac{\text{Im}(S_{ii})}{\text{Re}(S_{ii})}\right]. \tag{5.4.11a}$$

This line joins the center of the Γ_i-plane to points $\Gamma_i = S_{ii}^*$ or $\Gamma_i = 1/S_{ii}$.

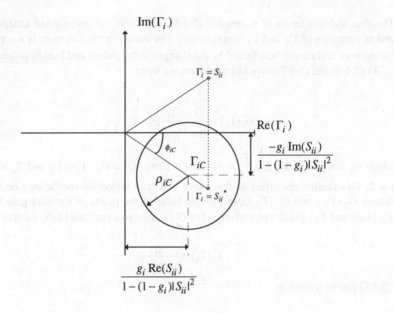

Fig. (5.4.1) $\Gamma_i -$ plane

The distant d_i of the center from the origin is

$$d_{iC} = \left| \frac{g_i |S_{ii}|}{1 - (1 - g_i)|S_{ii}|^2} \right| \qquad (5.4.11b)$$

and the radius of the circle is given as

$$\rho_{iC} = \left| \frac{(1 - |S_{ii}|^2)\sqrt{1 - g_i}}{1 - (1 - g_i)|S_{ii}|^2} \right|. \qquad (5.4.12)$$

A typical gain circle is shown in Fig. (5.4.1).

Unconditionally stable unilateral 2-port networks

The unilateral network is unconditionally stable if $|S_{ii}| < 1$. In this situation the maximum value of $G_i(\Gamma_i)$ is found for the conjugate matched condition when $\Gamma_i = S_{ii}^*$

and thus $g_i = 1$. Hence g_i can assume values between 0 and 1. A negative value for g_i gives a negative gain factor that is not possible with the present assumptions.

When $g_i = 0, d_{iC} = 0$ and $\rho_{iC} = 1$ and the circle of constant normalized power gain coincides with the circle of $|\Gamma_i| = 1$.

When $g_i = 1 - |S_{ii}|^2, |S_{ii}|^2 = 1 - g_i$ we have

$$\rho_{iC} = \frac{g_i \sqrt{(1 - g_i)}}{\left| 1 - (1 - g_i)^2 \right|}$$

and

$$d_{iC} = \frac{g_i \sqrt{(1 - g_i)}}{\left| 1 - (1 - g_i)^2 \right|}$$

and hence $\rho_i = d_i$. The center of the circle of constant gain $G_i(\Gamma_i)$, for $g_i = 1 - |S_{ii}|^2$, therefore, passes through the origin of Γ_i-plane. This is the circle of unity gain factor, inside this circle the gain factor $G_i(\Gamma_i)$ is greater than unity and outside less than unity.

When $g_i = 1$, $\Gamma_{iC} = S_{11}^*$ or $d_{iC} = |S_{ii}|$ and $\rho_{iC} = 0$. Hence for $g_i = 1$, the circle of constant normalized gain reduces to a point. This point is always inside the circle $|\Gamma_i| = 1$. It moves away from the center of Γ_i-plane as $|S_{ii}|$ increases, approaching $|\Gamma_i| = 1$ circle as it approaches unity. At this point, where $g_i = 1$, the source or load impedances are conjugate matched to the input or output impedances and the power associated with the port is a maximum. Typical circles of constant g_i are given in Fig. (5.4.2) for unconditionally stable cases and for increasing values of S_{ii}.

Potentially unstable unilateral 2-port networks

If the network is potentially unstable at either port, we have $|S_{ii}| > 1$ for the potentially unstable port. In this case as $1 - |S_{ii}|$ is negative and g_i is negative, as a positive value for g_i results in a negative transducer power gain. In this case, as we have seen, the conjugate matched condition is not possible. The transducer power gain can hence change from zero to infinity, as g_i changes from zero to minus infinity.

The circle for $g_i = 0$ again coincides with the circle of $|\Gamma_i| = 1$. For $g_i \to \infty$, $\Gamma_i \to 1 / S_{ii}$ and $\rho_i \to 0$, and hence the power gain circle reduces to a point inside the $|\Gamma_i| = 1$ circle. This point moves towards the center of the $|\Gamma_i| = 1$ circle as $|S_{ii}|$ increases. Typical circles of constant g_i are given in Fig. (5.4.3) for potentially unstable cases and for increasing values of S_{ii}.

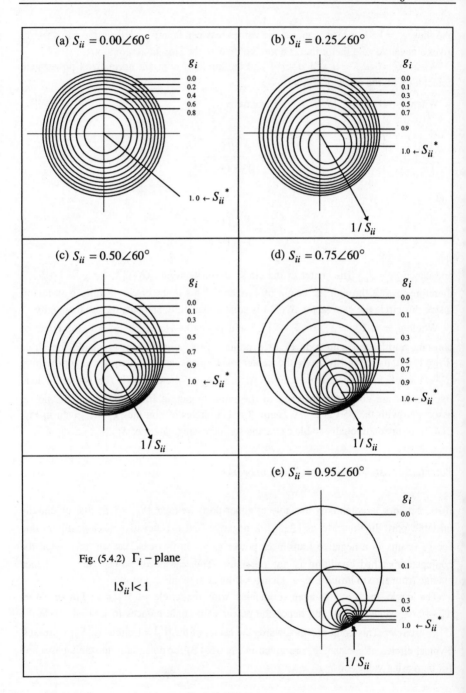

Fig. (5.4.2) Γ_i – plane

$|S_{ii}| < 1$

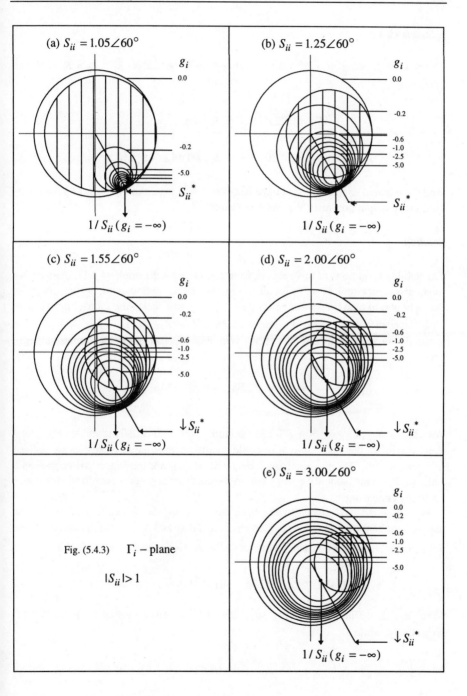

Fig. (5.4.3) Γ_i – plane

$|S_{ii}| > 1$

Example (5.4.1)

The scattering parameters of a 2-port network in a system of $\hat{Z}_1 = \hat{Z}_2 = R_0 = 50$ are given as

$$S_{11} = 0.72\angle 32°, \qquad S_{12} \approx 0,$$

$$S_{21} = 3.0\angle 0, \qquad S_{22} = 1.46\angle 64.8°.$$

Find the required source and load impedances to give a maximum transducer power gain, with a total output port resistance not less than 50.

Solution

The stability conditions of this network were considered in Example (4.4.1). As we have found, the input circuit is stable for all source impedances. For maximum contribution of the input port to total transducer power gain, we conjugate match the source impedance to input impedance of the network. Hence we have $Z_S = Z_{IN}^*$ or $\Gamma_S = \Gamma_{IN}^* = S_{11}^*$. The required source impedance is $81 - j128$. With this impedance, the input contribution to gain is

$$\frac{1}{1 - |S_{11}|^2} = 2.08 \quad \text{or} \quad 3.17\text{dB}.$$

With $\Gamma_{OT} = S_{22}$, $Z_{OT} = -30 + j70$ and the output port is potentially unstable. For a total resistance of 50, the load impedance should contribute a resistance of 80 or a normalized resistance of 1.6. For maximum gain, the circle of constant transducer power gain, in a complex Γ_L-plane, should be tangential to the circle of constant normalized resistance 1.6 in the same plane.

With d_{LC} and ρ_{LC} as the distance from the origin and the radius, respectively, of the required constant transducer power gain circle, and d_{Cr} and ρ_{Cr} as the distance from the origin and radius of the constant resistance circle, the two circles are tangential if

$$(\rho_{LC} + \rho_{Cr})^2 = d_{Cr}^2 + d_{LC}^2 - 2d_{Cr}d_{LC}\cos\phi_{LC}$$

where ϕ_{LC} is defined in (5.4.11a). d_{LC} and ρ_{LC} are given by (5.4.11b) and (5.4.12) respectively, while

$$d_{Cr} = \frac{r}{1+r}$$

and

$$\rho_{Cr} = \left| \frac{1}{1+r} \right|$$

as found in Section (3.2).

Substituting for d_{LC}, ρ_{LC}, d_{Cr}, and ρ_{Cr}, we have

$$A^2 g_L^2 - 4 g_L (A - 1) = 0$$

where

$$A = 1 + r \frac{1 + |S_{22}|^2}{1 - |S_{22}|^2} - 2r \frac{|S_{22}|}{1 - |S_{22}|^2} \cos \phi_{LC}.$$

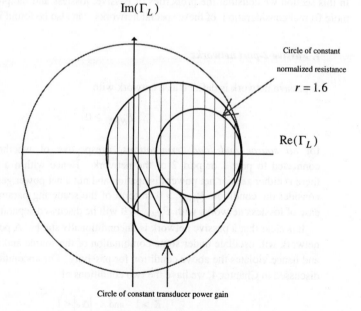

Γ_L – plane   shaded area is stable

The above quadratic equation has a trivial solution for $g_L = 0$ or a minimum gain with reactive loads, and a solution for maximum gain with

$$g_L = \frac{4(A-1)}{A^2}.$$

Substituting for $|S_{22}|$, r and ϕ_{LC}, we find $A = -1.67$, $g_L = -3.83$ and from (5.4.11b) $d_{LC} = 0.60$ and $\rho_{LC} = 0.268$.

The gain contribution from the output port is $g_L / (1 - |S_{22}|^2) = 3.38$ (5.29 dB) and $\Gamma_L = 0.51\angle - 38.6°$ for the required load impedance of $Z_L = 80 - j68.7$. With $10 \log |S_{21}|^2 = 9.54$, the total transducer power gain is

$$G_T = 3.17 + 9.54 + 5.29 = 18 \text{ dB}.$$

5.5 Power Consideration of Special 2-port Networks

In this section we consider the properties of passive, lossless and reciprocal networks. A more formal consideration of these special networks can also be found in Chapter 6.

1. Passive 2-port networks

A passive network is defined as a network with

$$P_{NWK} \geq 0 \qquad (5.5.1)$$

for *all source and load terminations*, irrespective of whether the source is connected to port 1 or port 2 of the network. Hence within a passive network, there is either zero or net power dissipation and not a net power generation. We first consider the condition $P_{NWK} > 0$ in term of the scattering parameters, the special case of lossless networks with $P_{NWK} = 0$ will be discussed separately.

It is clear that a passive network is unconditionally stable. A potentially unstable network will oscillate under some combination of the source and load impedances, and hence violates the above condition for passivity. For unconditional stability, as discussed in Chapter 4, we have the two conditions of

$$K > 1 \quad \text{and} \quad |\Delta_S| < 1 \qquad (5.5.2)$$

which are, therefore, necessary conditions for a network to be passive.

Condition (5.5.1) for passivity can be written as

$$P_{NWK} = P_{IN} - P_L = P_{IN}(1 - G_P) > 0.$$

If condition (5.5.2) is assumed both G_P and P_{IN} are positive, and for the network to be passive we require

$$G_P < 1 \qquad (5.5.3)$$

for all source and load impedances.

For the operating power gain to be less than unity for all load and source impedances, it should be less than unity with the network terminated by the impedances required for the conjugate matching. With this condition, and assuming that the source is connected to port 1, the operating power gain is a maximum and is given by (5.2.4b) as

$$G_{MAX} = \frac{|S_{21}|}{|S_{12}|}\left[K - \sqrt{(K^2 - 1)}\right]$$

where $K > 1$ as required by (5.5.2).

For the 2-port network to be passive, therefore, we require

$$\frac{|S_{21}|}{|S_{12}|}\left[K - \sqrt{(K^2 - 1)}\right] < 1$$

or

$$K - \frac{|S_{12}|}{|S_{21}|} < \sqrt{(K^2 - 1)}. \qquad (5.5.4a)$$

Similarly if the source is connected to port 2 of the network

$$K - \frac{|S_{21}|}{|S_{12}|} < \sqrt{(K^2 - 1)}. \qquad (5.5.4b)$$

Without loss of generality we can assume that $|S_{21}| > |S_{12}|$ and hence for the network to be passive (5.5.4a) should hold. With this assumption, both sides of (5.5.4a) are positive $(K > 1)$ and hence by squaring both sides we can write

$$2|S_{12}||S_{21}|K - |S_{12}|^2 - |S_{21}|^2 > 0.$$

Substituting for K as

$$K = \frac{1 - |S_{11}|^2 - |S_{22}|^2 + |\Delta_S|^2}{2|S_{12}S_{21}|},$$

we finally have

$$|S_{11}|^2 + |S_{21}|^2 + |S_{22}|^2 + |S_{12}|^2 < 1 + |\Delta_S|^2. \tag{5.5.5}$$

The inequality (5.5.5) together with conditions (5.5.2) are the *necessary and sufficient* conditions for the network to be passive. After expansion of $|\Delta_S|^2$ and some rearrangement, (5.5.5) can be written as

$$(1 - |S_{11}|^2 - |S_{21}|^2)(1 - |S_{22}|^2 - |S_{12}|^2) >$$

$$(S_{11}^*S_{12} + S_{22}S_{21}^*)(S_{11}S_{12}^* + S_{22}^*S_{21}) \tag{5.5.6}$$

with right-hand side of (5.5.6) always positive, the two left-hand terms have to be either positive or negative. If these terms are both negative, then the sum of two terms is also negative and hence

$$|S_{11}|^2 + |S_{22}|^2 + |S_{21}|^2 + |S_{12}|^2 > 2$$

which is contrary to (5.5.5) ($|\Delta_S| < 1$) for the network to be passive. Hence the following two relations are also true for any passive network, and together with (5.5.2) can be considered as necessary and sufficient conditions for a circuit to be passive:

$$1 - |S_{11}|^2 - |S_{21}|^2 > 0, \tag{5.5.7a}$$

$$1 - |S_{22}|^2 - |S_{12}|^2 > 0. \tag{5.5.7b}$$

2. Lossless 2-port network

If a 2-port network is lossless

$$P_{NWK} = 0$$

for all load impedances. P_{NWK} is given by (5.1.17) and hence we have

$$P_{NWK} = e_S^2 \frac{1 - |\Gamma_S|^2}{|1 - \Gamma_S\Gamma_{IN}|^2}\left[1 - \frac{|S_{11} - \Delta_S\Gamma_L|^2}{|1 - \Gamma_L S_{22}|^2} - \frac{|S_{21}|^2(1 - |\Gamma_L|^2)}{|1 - \Gamma_L S_{22}|^2}\right] = 0$$

or

$$\left|1 - \Gamma_L S_{22}\right|^2 - \left|S_{11} - \Delta_S \Gamma_L\right|^2 - \left|S_{21}\right|^2 (1 - \left|\Gamma_L\right|^2) = 0. \qquad (5.5.8)$$

For the above relation to be satisfied for all possible Γ_L we require the following identities to hold

$$1 - \left|S_{11}\right|^2 - \left|S_{21}\right|^2 = 0, \qquad (5.5.9a)$$

$$\left|S_{21}\right|^2 + \left|S_{22}\right|^2 - \left|\Delta_S\right|^2 = 0, \qquad (5.5.9b)$$

$$S_{22} - S_{11}^* \Delta_S = 0. \qquad (5.5.9c)$$

The above relations are found by expanding (5.5.8) and setting the constant term and the coefficients of all terms containing Γ_L, Γ_L^* and $\Gamma_L \Gamma_L^*$ equal to zero.

Changing the position of the source and the load, we similarly have

$$1 - \left|S_{22}\right|^2 - \left|S_{12}\right|^2 = 0, \qquad (5.5.10a)$$

$$\left|S_{12}\right|^2 + \left|S_{11}\right|^2 - \left|\Delta_S\right|^2 = 0, \qquad (5.5.10b)$$

$$S_{11} - S_{22}^* \Delta_S = 0. \qquad (5.5.10c)$$

Multiplying both sides of (5.5.9c) and (5.5.10c) by S_{22}^* and S_{11}^* respectively and subtracting we have

$$S_{22} S_{22}^* - S_{11} S_{11}^* = 0$$

which together with (5.5.9a) and (5.5.10a) lead to the following new identities:

$$\left|S_{11}\right| = \left|S_{22}\right| \qquad (5.5.11a)$$

and

$$\left|S_{12}\right| = \left|S_{21}\right|. \qquad (5.5.11b)$$

In view of the above relations, (5.5.9a) and (5.5.10a) can be written as

$$|S_{11}|^2 + |S_{12}|^2 = 1,$$ (5.5.12a)

$$|S_{22}|^2 + |S_{21}|^2 = 1.$$ (5.5.12b)

Adding the above equations

$$|S_{11}|^2 + |S_{12}|^2 + |S_{21}|^2 + |S_{22}|^2 = 2$$ (5.5.13)

From (5.5.9a) and (5.5.10b) or from (5.5.10a) and (5.5.9b), we can easily verify that

$$|\Delta_S| = 1$$ (5.5.14)

and

$$K = \frac{1 - |S_{11}|^2 - |S_{22}|^2 + |\Delta_S|^2}{2|S_{12}||S_{21}|} = \frac{|S_{21}|^2 + |S_{12}|^2}{2|S_{12}||S_{21}|} = 1.$$ (5.5.15)

It can also be shown that unless $|S_{11}| = |S_{22}| = 0$ or $|S_{11}| = |S_{22}| = 1$, then for $|\Delta_S| = 1$ to be satisfied we require

$$(\phi_{11} - \phi_{21}) + (\phi_{22} - \phi_{12}) = \pi$$ (5.5.16)

where ϕ_{11}, ϕ_{12}, ϕ_{21} and ϕ_{22} are the phases of S_{11}, S_{12}, S_{21} and S_{22} respectively.

Although for a lossless network all the conditions (5.5.10) to (5.5.16) hold, the conditions (5.5.14) and (5.5.15) are the *necessary and sufficient* conditions for the network to be lossless.

3. Reciprocal 2-port networks

A reciprocal 2-port network can be defined as a network with the property $Z_{12} = Z_{21}$ or $Y_{12} = Y_{21}$. In Section (2.4) we have defined the Z-parameters and Y-parameters by relations between currents and voltages as

$$\begin{vmatrix} V_1 \\ V_2 \end{vmatrix} = \begin{vmatrix} Z_{11} & Z_{12} \\ Z_{21} & Z_{22} \end{vmatrix} \begin{vmatrix} I_1 \\ I_2 \end{vmatrix}$$ (5.5.17)

and

$$\begin{vmatrix} I_1 \\ I_2 \end{vmatrix} = \begin{vmatrix} Y_{11} & Y_{12} \\ Y_{21} & Y_{22} \end{vmatrix} \begin{vmatrix} V_1 \\ V_2 \end{vmatrix} \tag{5.5.18}$$

respectively.

In Equation (5.5.17) if port 1 is open-circuited and port 2 is connected to a current generator (case a) $I_1^{(a)} = 0$, then $V_1^{(a)} = Z_{12} I_2^{(a)}$. Similarly if port-2 is open-circuited and port 1 is connected to a current generator (case b), $I_2^{(b)} = 0$ and $V_2^{(b)} = Z_{21} I_1^{(b)}$. Hence with $Z_{12} = Z_{21}$, if $I_1^{(b)} = I_2^{(a)}$ then $V_1^{(a)} = V_2^{(b)}$. This is shown in Fig. (5.5.1). Hence for a reciprocal 2-port network, if equal currents are injected in turn to each port, equal voltages are obtained across the remaining open-circuited port.

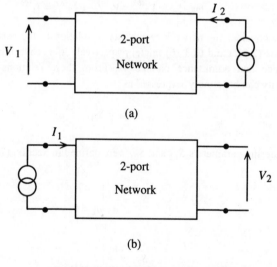

(a)

(b)

Fig. (5.5.1) $I_2^{(a)} = I_1^{(b)} \rightarrow V_2^{(b)} = V_1^{(a)}$

Similarly from Equation (5.5.18), if $V_1^{(a)} = 0$ or port 1 is short circuited $I_1^{(a)} = Y_{12} V_2^{(a)}$ and if port 2 is short-circuited $V_2^{(b)} = 0$ and $I_2^{(b)} = Y_{21} V_1^{(b)}$. Hence with $Y_{12} = Y_{21}$ and $V_1^b = V_2^a$, $I_1^{(a)} = I_2^{(b)}$. The situation is shown in Fig. (5.5.2). Hence for a reciprocal 2-port network, if equal voltages are applied in turn across each port, equal currents will flow through the other short-circuited port.

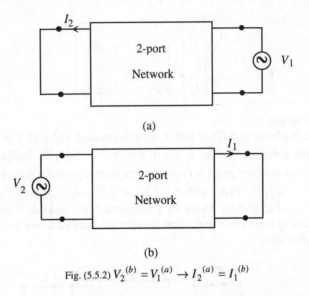

(a)

(b)

Fig. (5.5.2) $V_2^{(b)} = V_1^{(a)} \rightarrow I_2^{(a)} = I_1^{(b)}$

We now consider the network reciprocity in relation to the scattering parameters. Equations (2.3.7) and (2.3.16) relate, respectively, the scattering parameters to the impedance and admittance parameters. From these relations the condition of reciprocity can be simply expressed as

$$S_{12} = S_{21}. \tag{5.5.19}$$

The scattering parameters S_{12} and S_{21} were defined in Section (2.3) as

$$S_{12} = \frac{b_1}{a_2}\bigg|_{a_1 = 0},$$

$$S_{21} = \frac{b_2}{a_1}\bigg|_{a_2 = 0}.$$

As we have discussed in the same section, $a_1 = 0$ when port 1 is terminated by impedance R_0 and similarly $a_2 = 0$ when port 2 is terminated by the same impedance. Hence for a reciprocal network, the reflected wave from port 1 due to an incident wave on port 2, when port 1 is terminated by its reference impedance, is equal to the reflected wave from port 2 due to an equal incident wave on port 1, when port 2 is terminated by its reference impedance. This is shown in Fig. (5.5.3).

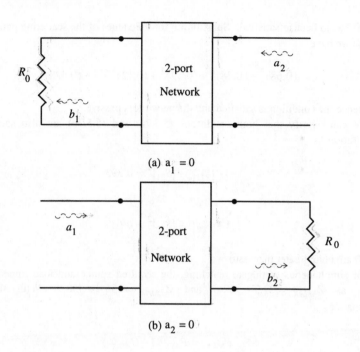

(a) $a_1 = 0$

(b) $a_2 = 0$

Fig. (5.5.3) $a_2{}^{(a)} = a_1{}^{(b)} \rightarrow b_1{}^{(a)} = b_2{}^{(b)}$

Example (5.5.1)

The scattering parameters of a 2-port network are given by

$$S_{11} = 0.68\angle 74°, \qquad S_{12} = 0.12\angle 32°,$$

$$S_{21} = 0.38\angle -63°, \qquad S_{22} = 0.59\angle 18°.$$

Show that the network satisfies the conditions stated for a passive network. If the power available from the source is 5, find P_{NWK} with the input and output ports simultaneously matched.

Solution

For the above network $K = 4.08$ and $\Delta_S = 0.428\angle 97.1°$ or $|\Delta_S| = 0.428$. Hence, $K > 1$ and $|\Delta_S| < 1$ and the network is unconditionally stable. For the network to be passive

(5.5.5) has to be also satisfied. Substituting for the values of the scattering parameters in (5.5.5) we have

$$(0.68)^2 + (0.38)^2 + (0.59)^2 + (0.12)^2 < 1 + (0.428)^2$$

and hence the condition is satisfied and the network is passive.

We can verify that both conditions (5.5.7a) and (5.5.7b) are also satisfied. By substitution

and

$$1 - |S_{11}|^2 - |S_{21}|^2 = 0.393$$

$$1 - |S_{22}|^2 - |S_{12}|^2 = 0.637$$

which are both greater than zero.

For simultaneous conjugate matching, the required source and load impedances are given as $Z_{SMAX} = 27.7 - j62.3$ and $Z_{LMAX} = 156.4 - j58.9$. With the above terminations

$$P_{NWK} = P_{AVS} (1 - G_{MAX})$$

and

$$G_{MAX} = \frac{|S_{21}|}{|S_{12}|} \left[K - \sqrt{(K^2 - 1)} \right].$$

Hence $G_{MAX} = 0.394$ and $P_{NWK} = 5(1 - 0.394) = 3.03$. As expected a net power is delivered to the network and the network is dissipative.

Example (5.5.2)

Show that a network with the given scattering parameters

$$S_{11} = 0.6 \angle 68°, \qquad S_{12} = 0.8 \angle -42°,$$

$$S_{21} = 0.8 \angle 10°, \qquad S_{22} = 0.6 \angle 80°$$

is lossless. If the network is connected to a source of available power 5, show that the maximum available power at the load is also equal to 5.

Solution

We have

$$\Delta_S = S_{11}S_{22} - S_{12}S_{21} = 0.36\angle148° - 0.64\angle-32° = -0.848 + j0.530$$

or

$$|\Delta_S| = 1$$

and also

$$K = \frac{1 = 0.36 - 0.36 + 1}{2 \times 0.8 \times 0.8} = 1$$

In addition we have

$$|S_{11}|^2 + |S_{21}|^2 + |S_{22}|^2 + |S_{12}|^2 = 1 + |\Delta_S|^2 = 2$$

and hence the network is lossless.

The network being lossless, it also satisfies the following conditions

$$|S_{11}| = |S_{22}| \, (= 0.6)$$

$$|S_{12}| = |S_{21}| \, (= 0.8)$$

as well as

$$(\phi_{11} - \phi_{21}) + (\phi_{22} - \phi_{12}) \, (= 58° + 122°) = 180°.$$

Maximum available power at the load is given by

$$P_{MAX} = P_{AVS} \frac{|S_{21}|}{|S_{12}|} \left[K - \sqrt{(K^2 - 1)} \right] = P_{AVS} \, (K = 1, |S_{21}| = |S_{12}|)$$

or this power is also 5 being equal to the power available from the source.

Example (5.5.3)

Show that the network with the given scattering parameters

$$S_{11} = 0.6\angle32°, \qquad S_{12} = 0.4\angle102°,$$

$$S_{22} = 0.4\angle102°, \qquad S_{22} = 0.8\angle44°.$$

measured in a system of reference impedance $R_0 = 50$ is a reciprocal network. Verify that $Z_{12} = Z_{21}$.

Solution

With

$$S_{12} = S_{21} \ (= 0.4\angle 102°),$$

the network is reciprocal. Using the conversion of scattering parameters to the impedance parameters as given by (2.3.8b) and (2.3.8c) we have $Z_{12} = Z_{21} \ (= 103.55\angle 164.6°)$ as required.

Chapter 6

Generalized Scattering Parameters of a 2-port Network

Many expressions derived in the previous chapters can be written in a more concise manner, if the scattering parameters are given in systems of specific reference impedances. These reference impedances do not have to be necessarily resistive. Specifically, a very convenient set of expressions for the power gains can be found, if the reference impedances for the input and output ports are made equal to that of the source and load impedances, respectively. The scattering parameters defined with two general reference impedances are known as 'the generalized scattering parameters'. The measurement of scattering parameters for each different set of reference impedances is not practical. Hence the new sets of scattering parameters are calculated from the parameters measured in systems of resistive references, with the conversion expressions given.

6.1 The Generalized Scattering Matrix of a 2-port Network

As we have discussed previously, the scattering parameters are defined in terms of a set of reference impedances or reference reflection coefficients. In the case of a 2-port network we require two reference impedances, one assigned to each port. Designating these impedances as $\hat{Z}_1$ and $\hat{Z}_2$, the reference reflection coefficients γ_1 and γ_2 are related to these impedances as

$$\gamma_1 = \frac{\hat{Z}_1 - R_{01}}{\hat{Z}_1 + R_{01}}, \qquad \gamma_2 = \frac{\hat{Z}_2 - R_{02}}{\hat{Z}_2 + R_{02}},$$

where R_{01} and R_{02} are the resistive references for the measured scattering parameters. In the previous sections we assumed that $R_{01} = R_{02} = R_0$ or the reference impedance for the two ports were equal. This assumption needs not to be made here.

Current and voltage scattering matrices

To define the scattering parameters of a 2-port network, in a system of reference impedances $\hat{Z}_1$ and $\hat{Z}_2$, we connect the network to two generators with assumed internal impedances equal to $\hat{Z}_1$ and $\hat{Z}_2$ as shown in Fig. (6.1.1).

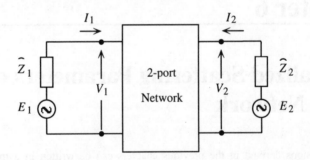

Fig. (6.1.1)

As before, the generator emf of port 1 can be denoted by E_1, the current flowing into port 1 by I_1 and the voltage across port 1 by V_1. The parameters relating to port 2 can similarly be defined, but with the subscripts changing from 1 to 2.

We define the incident currents

$$\mathbf{I}^i(\gamma_1, \gamma_2) = \begin{bmatrix} I_1^i(\gamma_1) \\ I_2^i(\gamma_2) \end{bmatrix}$$

and voltages

$$\mathbf{V}^i(\gamma_1,\gamma_2) = \begin{bmatrix} V_1^i(\gamma_1) \\ V_2^i(\gamma_2) \end{bmatrix}$$

as the currents through and voltages across the two ports, when the 2-port network is disconnected and each generator is terminated by the *conjugate* of its reference impedance, as in Fig. (6.1.2).

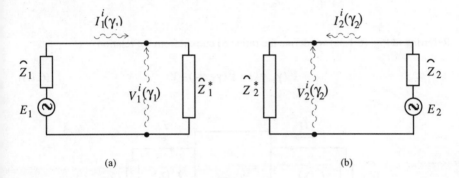

(a)   (b)

Fig. (6.1.2)

Hence by referring to Figs. (6.1.1) (6.1.2) we can write

$$\mathbf{I}^i(\gamma_1,\gamma_2) = \begin{bmatrix} \dfrac{E_1}{2\widehat{R}_1} \\[2mm] \dfrac{E_2}{2\widehat{R}_2} \end{bmatrix} = \begin{bmatrix} \dfrac{V_1 + \widehat{Z}_1 I_1}{2\widehat{R}_1} \\[2mm] \dfrac{V_2 + \widehat{Z}_2 I_2}{2\widehat{R}_2} \end{bmatrix} \qquad (6.1.1)$$

where $\widehat{R}_1 = \dfrac{1}{2}(\widehat{Z}_1 + \widehat{Z}_1^*) = \mathrm{Re}(\widehat{Z}_1)$ and $\widehat{R}_2 = \dfrac{1}{2}(\widehat{Z}_2 + \widehat{Z}_2^*) = \mathrm{Re}(\widehat{Z}_2)$. Similarly

$$\mathbf{V}^i(\gamma_1,\gamma_2) = \begin{bmatrix} \dfrac{\widehat{Z}_1^* E_1}{2\widehat{R}_1} \\[2mm] \dfrac{\widehat{Z}_2^* E_2}{2\widehat{R}_2} \end{bmatrix} = \begin{bmatrix} \dfrac{\widehat{Z}_1^*(V_1 + \widehat{Z}_1 I_1)}{2\widehat{R}_1} \\[2mm] \dfrac{\widehat{Z}_2^*(V_2 + \widehat{Z}_2 I_2)}{2\widehat{R}_2} \end{bmatrix} \qquad (6.1.2)$$

From (6.1.1) and (6.1.2)

$$\mathbf{V}^i(\gamma_1,\gamma_2) = \hat{\mathbf{Z}}^*\mathbf{I}^i(\gamma_1,\gamma_2) \tag{6.1.3}$$

where

$$\hat{\mathbf{Z}} = \begin{bmatrix} \hat{Z}_1 & 0 \\ 0 & \hat{Z}_2 \end{bmatrix} \quad \text{and} \quad \hat{\mathbf{Z}}^* = \begin{bmatrix} \hat{Z}_1^* & 0 \\ 0 & \hat{Z}_2^* \end{bmatrix}.$$

Referring to Fig. (6.1.3a) we define the reflected current from the relation

$$\mathbf{I}^r(\gamma_1,\gamma_2) = \mathbf{I}^i(\gamma_1,\gamma_2) - \mathbf{I}. \tag{6.1.4}$$

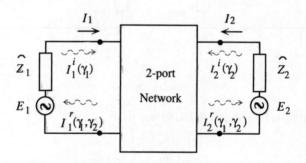

Fig. (6.1.3a)

Hence in terms of the total currents and voltages

$$\mathbf{I}^r(\gamma_1,\gamma_2) \equiv \begin{bmatrix} I_1^r(\gamma_1,\gamma_2) \\ \\ I_2^r(\gamma_1,\gamma_2) \end{bmatrix} = \begin{bmatrix} \dfrac{E_1}{2\hat{R}_1} - I_1 \\ \\ \dfrac{E_2}{2\hat{R}_2} - I_2 \end{bmatrix} = \begin{bmatrix} \dfrac{V_1 - \hat{Z}_1^* I_1}{2\hat{R}_1} \\ \\ \dfrac{V_2 - \hat{Z}_2^* I_2}{2\hat{R}_2} \end{bmatrix}. \tag{6.1.5}$$

Similarly referring to Fig. (6.1.3b) we define the reflected voltage by the relation

$$\mathbf{V}^r(\gamma_1,\gamma_2) = \mathbf{V} - \mathbf{V}^i(\gamma_1,\gamma_2). \tag{6.1.6}$$

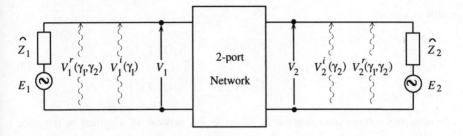

Fig. (6.1.3b)

Hence again in terms of currents and voltages

$$\mathbf{V}^r(\gamma_1,\gamma_2) = \begin{bmatrix} V_1 - \dfrac{\hat{Z}_1^* E_1}{2\hat{R}_1} \\[2ex] V_2 - \dfrac{\hat{Z}_2^* E_2}{2\hat{R}_2} \end{bmatrix} = \begin{bmatrix} \dfrac{\hat{Z}_1(V_1 - \hat{Z}_1^* I_1)}{2\hat{R}_1} \\[2ex] \dfrac{\hat{Z}_2(V_2 - \hat{Z}_2^* I_2)}{2\hat{R}_2} \end{bmatrix}. \tag{6.1.7}$$

From (6.1.5) and (6.1.7)

$$\mathbf{V}^r(\gamma_1,\gamma_2) = \hat{\mathbf{Z}}\mathbf{I}^r(\gamma_1,\gamma_2). \tag{6.1.8}$$

The relation between $\mathbf{I}^r(\gamma_1,\gamma_2)$ and $\mathbf{I}^i(\gamma_1,\gamma_2)$, however, can be written as

$$\mathbf{I}^r(\gamma_1,\gamma_2) = \mathbf{S}^I(\gamma_1,\gamma_2)\mathbf{I}^i(\gamma_1,\gamma_2) \tag{6.1.9}$$

where

$$\mathbf{S}^I(\gamma_1,\gamma_2) = \begin{bmatrix} S_{11}^I(\gamma_1,\gamma_2) & S_{12}^I(\gamma_1,\gamma_2) \\[2ex] S_{21}^I(\gamma_1,\gamma_2) & S_{22}^I(\gamma_1,\gamma_2) \end{bmatrix} \tag{6.1.10}$$

defines the current scattering matrix of the 2-port network in a system of reference impedances $\hat{Z}_1$ and $\hat{Z}_2$ (or reference coefficients γ_1 and γ_2).

Similarly

$$\mathbf{V}^r(\gamma_1,\gamma_2) = \mathbf{S}^V(\gamma_1,\gamma_2)\mathbf{V}^i(\gamma_1,\gamma_2) \tag{6.1.11}$$

with

$$\mathbf{S}^V(\gamma_1,\gamma_2) = \begin{bmatrix} S_{11}^V(\gamma_1,\gamma_2) & S_{12}^V(\gamma_1,\gamma_2) \\[2mm] S_{21}^V(\gamma_1,\gamma_2) & S_{22}^V(\gamma_1,\gamma_2) \end{bmatrix} \tag{6.1.12}$$

defining the voltage scattering matrix of the 2-port network in a system of reference impedances $\hat{Z}_1$ and $\hat{Z}_2$.

Wave amplitude scattering matrix

The incident wave amplitudes are related to the incident currents and voltages and are defined as

$$\mathbf{a}(\gamma_1,\gamma_2) \equiv \begin{bmatrix} a_1(\gamma_1) \\[1mm] a_2(\gamma_2) \end{bmatrix} = \begin{bmatrix} \sqrt{\hat{R}_1} & 0 \\[1mm] 0 & \sqrt{\hat{R}_2} \end{bmatrix} \mathbf{I}^i(\gamma_1,\gamma_2)$$

$$= \begin{bmatrix} \sqrt{\hat{R}_1}/\hat{Z}_1^* & 0 \\[1mm] 0 & \sqrt{\hat{R}_2}/\hat{Z}_2^* \end{bmatrix} \mathbf{V}^i(\gamma_1,\gamma_2). \tag{6.1.13}$$

Similarly the reflected wave amplitudes are defined as

$$\mathbf{b}(\gamma_1,\gamma_2) \equiv \begin{bmatrix} b_1(\gamma_1,\gamma_2) \\[1mm] b_2(\gamma_1,\gamma_2) \end{bmatrix} = \begin{bmatrix} \sqrt{\hat{R}_1} & 0 \\[1mm] 0 & \sqrt{\hat{R}_2} \end{bmatrix} \mathbf{I}^r(\gamma_1,\gamma_2)$$

$$= \begin{bmatrix} \sqrt{\hat{R}_1}/\hat{Z}_1 & 0 \\[1mm] 0 & \sqrt{\hat{R}_2}/\hat{Z}_2 \end{bmatrix} \mathbf{V}^r(\gamma_1,\gamma_2). \tag{6.1.14}$$

The incident and reflected wave amplitudes can be expressed in terms of the total currents and voltages as

$$\mathbf{a}(\gamma_1,\gamma_2) = \frac{1}{2} \begin{bmatrix} 1/\sqrt{\hat{R}_1} & 0 \\[2mm] 0 & 1/\sqrt{\hat{R}_2} \end{bmatrix} \{\mathbf{V} + \hat{\mathbf{Z}}\mathbf{I}\} \tag{6.1.15}$$

and

$$\mathbf{b}(\gamma_1,\gamma_2) = \frac{1}{2} \begin{bmatrix} 1/\sqrt{\hat{R}_1} & 0 \\ 0 & 1/\sqrt{\hat{R}_2} \end{bmatrix} \left\{ \mathbf{V} - \hat{\mathbf{Z}}^*\mathbf{I} \right\}. \tag{6.1.16}$$

The matrix relation

$$\mathbf{b}(\gamma_1,\gamma_2) = \mathbf{S}(\gamma_1,\gamma_2)\mathbf{a}(\gamma_1,\gamma_2) \tag{6.1.17}$$

where

$$\mathbf{S}(\gamma_1,\gamma_2) \equiv \begin{bmatrix} S_{11}(\gamma_1,\gamma_2) & S_{12}(\gamma_1,\gamma_2) \\ S_{21}(\gamma_1,\gamma_2) & S_{22}(\gamma_1,\gamma_2) \end{bmatrix} \tag{6.1.18}$$

defines the wave amplitude scattering matrix of the network.

$S_{11}(\gamma_1,\gamma_2)$, $S_{12}(\gamma_1,\gamma_2)$, $S_{21}(\gamma_1,\gamma_2)$ and $S_{22}(\gamma_1,\gamma_2)$ are known as the 2-port wave amplitude generalized scattering parameters, or simply as the generalized scattering parameters of the 2-port network. Generalized scattering parameters can be defined in words as:

(1) $S_{11}(\gamma_1,\gamma_2)$ is the ratio of the reflected wave $b_1(\gamma_1,\gamma_2)$ from port 1 to the incident wave $a_1(\gamma_1)$ on port 1 when source 1, connected in series with $\hat{Z}_1$, is generating and port 2 is terminated by $\hat{Z}_2$.

(2) $S_{12}(\gamma_1,\gamma_2)$ is the ratio of the reflected wave $b_1(\gamma_1,\gamma_2)$ from port 1 to the incident wave $a_2(\gamma_2)$ on port 2 when source 2, connected in series with $\hat{Z}_2$, is generating and port 1 is terminated by $\hat{Z}_1$.

(3) $S_{21}(\gamma_1,\gamma_2)$ is the ratio of the reflected wave $b_2(\gamma_1,\gamma_2)$ from port 2 to the incident wave $a_1(\gamma_1)$ on port 1 when source 1, connected in series with $\hat{Z}_1$, is generating and port 2 is terminated by $\hat{Z}_2$.

(4) $S_{22}(\gamma_1,\gamma_2)$ is the ratio of the reflected wave $b_2(\gamma_1,\gamma_2)$ from port 2 to the incident wave $a_2(\gamma_2)$ on port 2 when source 2, connected in series with $\hat{Z}_2$, is generating and port 1 is terminated by $\hat{Z}_1$.

When $\hat{Z}_1 = R_{01}, \gamma_1 = 0$ and similarly when $\hat{Z}_2 = R_{02}, \gamma_2 = 0$. Hence $\mathbf{S}(0,0)$ is the scattering matrix in a resistive reference system. While measurements are made in a resistive reference system, as we shall see in Section (6.3), the scattering matrix in any other system of reference impedance $\hat{Z}_1$ and $\hat{Z}_2$ (or equivalently reference reflection coefficients γ_1 and γ_2) can readily be derived from the measured values. For convenience we denote $\mathbf{S}(0,0)$ as $\mathbf{S}$. Hence S_{11}, S_{12}, S_{21} and S_{22} are the measured scattering parameters, in a system of standard resistive references R_{01} and R_{02}. Similarly we denote $a_1(0), a_2(0), b_1(0,0)$ and $b_2(0,0)$ as a_1, a_2, b_1 and b_2 respectively.

Example (6.1.1)

From the definition of the scattering parameters, find the scattering parameters of a 2-port network consisting of a series resistance Z_a. Assume a generalized system of reference impedances $\hat{Z}_1$ and $\hat{Z}_2$.

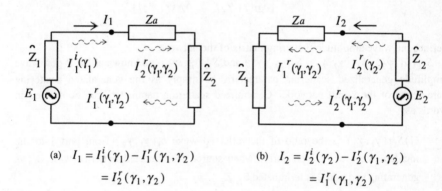

(a) $I_1 = I_1^i(\gamma_1) - I_1^r(\gamma_1, \gamma_2)$ (b) $I_2 = I_2^i(\gamma_2) - I_2^r(\gamma_1, \gamma_2)$

         $= I_2^r(\gamma_1, \gamma_2)$ $= I_1^r(\gamma_1, \gamma_2)$

Solution

In Fig. (a) a generator E_1 in series with the reference impedance $\hat{Z}_1$ and the series impedance Z_a, is terminated by the reference impedance $\hat{Z}_2$. The scattering parameter $S_{11}(\gamma_1, \gamma_2)$ can be found by referring to this figure and definition (1) of the scattering parameters.

The incident and reflected currents are given by

$$I_1^i(\gamma_1) = \frac{E_1}{2\hat{R}_1} \quad \text{and} \quad I_1^r(\gamma_1, \gamma_2) = \frac{(Z_a + \hat{Z}_2 - \hat{Z}_1^*)E_1}{2\hat{R}_1(\hat{Z}_1 + Z_a + \hat{Z}_2)}$$

and hence

$$a_1(\gamma_1) = \frac{E_1}{2\sqrt{\widehat{R}_1}} \quad \text{and} \quad b_1(\gamma_1, \gamma_2) = \frac{(Z_a + \widehat{Z}_2 - \widehat{Z}_1^*)E_1}{2\sqrt{\widehat{R}_1}(\widehat{Z}_1 + Z_a + \widehat{Z}_2)}$$

or

$$S_{11}(\gamma_1, \gamma_2) = \frac{b_1(\gamma_1, \gamma_2)}{a_1(\gamma_1)} = \frac{(\widehat{Z}_2 + Z_a) - \widehat{Z}_1^*}{\widehat{Z}_1 + (\widehat{Z}_2 + Z_a)} = \frac{Z - 2\widehat{R}_1}{Z}$$

where

$$Z = \widehat{Z}_1 + \widehat{Z}_2 + Z_a.$$

To find $S_{21}(\gamma_1, \gamma_2)$ we note that

$$I_2^r(\gamma_1, \gamma_2) = I_1^i(\gamma_1) - I_1^r(\gamma_1, \gamma_2) = I_1$$

and hence

$$b_2(\gamma_1, \gamma_2) = \frac{\sqrt{\widehat{R}_2}E_1}{(\widehat{Z}_1 + Z_a + \widehat{Z}_2)}$$

or from definition (3) for $S_{21}(\gamma_1, \gamma_2)$, we have

$$S_{21}(\gamma_1, \gamma_2) = \frac{2\sqrt{\widehat{R}_1}\sqrt{\widehat{R}_2}}{Z}$$

by referring to Fig. (b) and definitions (2) and (4), we can write $S_{12}(\gamma_1, \gamma_2)$ and $S_{22}(\gamma_1, \gamma_2)$ as

$$S_{12}(\gamma_1, \gamma_2) = \frac{2\sqrt{\widehat{R}_1}\sqrt{\widehat{R}_2}}{Z} \quad \text{and} \quad S_{22}(\gamma_1, \gamma_2) = \frac{Z - 2\widehat{R}_2}{Z}.$$

With $\widehat{Z}_1 = \widehat{Z}_2 = R_{01} = R_{02} = R_0$, we have

$$S_{11} = \frac{Z_a}{Z} \qquad S_{12} = \frac{2R_0}{Z}$$

$$S_{21} = \frac{2R_0}{Z} \qquad S_{22} = \frac{Z_a}{Z}$$

where $Z = 2R_0 + Z_a$ as in Example (2.1.1).

Example (6.1.2)

Find the scattering parameters of a series impedance $Z_a = 100 + j50$ for reference impedances (i) $\hat{Z}_1 = \hat{Z}_2 = 50$, (ii) $\hat{Z}_1 = 50 + j25$ and $\hat{Z}_2 = 50 - j25$, (iii) $\hat{Z}_1 = 50 + j25$ and $\hat{Z}_2 = 75 - j50$.

Solution

Here we use the expressions for the scattering parameters as found in the previous example.

(i) $Z = Z_a + \hat{Z}_1 + \hat{Z}_2 = 100 + j50 + 50 + 50 = 200 + j50$, $\quad \hat{R}_1 = \hat{R}_2 = 50$ and after substitution

$$S_{11}(\gamma_1, \gamma_2) = S_{22}(\gamma_1, \gamma_2) = 0.53 + j0.12,$$

$$S_{12}(\gamma_1, \gamma_2) = S_{21}(\gamma_1, \gamma_2) = 0.47 - j0.12.$$

(ii) $Z = Z_a + \hat{Z}_1 + \hat{Z}_2 = 100 + j50 + 50 + j25 + 50 - j25 = 200 + j50$, $\hat{R}_1 = \hat{R}_2 = 50$, hence for this case the scattering parameters will be identical with that of case (i).

(iii) $Z = Z_a + \hat{Z}_1 + \hat{Z}_2 = 100 + j50 + 50 + j25 + 75 - j50 = 225 + j25$, $\quad \hat{R}_1 = 50$, $\hat{R}_2 = 75$ and after substitution

$$S_{11}(\gamma_1, \gamma_2) = 0.56 + j0.05, \qquad S_{12}(\gamma_1, \gamma_2) = 0.54 - j0.06,$$

$$S_{21}(\gamma_1, \gamma_2) = 0.54 - j0.06, \qquad S_{22}(\gamma_1, \gamma_2) = 0.34 + j0.07.$$

Example (6.1.3)

From the definition of the scattering parameters, find the scattering parameters of a 2-port network consisting of a shunt admittance Y_b. Assume a generalized system of reference impedances $\hat{Z}_1$ and $\hat{Z}_2$.

Solution

The procedure in this case is similar to that of the last problem. For this case, however, if port 1 is generating, the reflected voltage from port 2 is equal to the sum of the incident

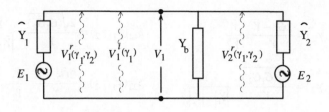

$$V_1 = V_1^i(\gamma_1) + V_1^r(\gamma_1, \gamma_2) = V_2^r(\gamma_1, \gamma_2)$$

and reflected voltages related to port 1. Similarly if port 2 is generating the reflected voltage from port 1 is equal to the sum of the incident and the reflected voltages related to port 2. Hence it is appropriate to start with the voltage scattering parameters. Subsequently we convert the voltage scattering parameters to the wave amplitude scattering parameters, using the given appropriate relations.

Referring to the above figure we can write

$$V_1 = V_1^i(\gamma_1, \gamma_2) + V_1^r(\gamma_1) = V_2^r(\gamma_1, \gamma_2),$$

and easily verify that with $\hat{Y}_1 = 1/\hat{Z}_1$, $\hat{Y}_2 = 1/\hat{Z}_2$, $\hat{G}_1 = \mathrm{Re}(\hat{Y}_1)$ and $\hat{G}_2 = \mathrm{Re}(\hat{Y}_2)$,

$$S_{11}^V(\gamma_1, \gamma_2) = \frac{\hat{Z}_1}{\hat{Z}_1^*} \times \frac{1/Y_b + \hat{Y}_2) - \hat{Z}_1^*}{\hat{Z}_1 + 1/(Y_b + \hat{Y}_2)} = \frac{2\hat{G}_1 - Y}{Y}$$

and

$$S_{21}^V(\gamma_1, \gamma_2) = 1 + \frac{2\hat{G}_1 - Y}{Y} = \frac{2\hat{G}_1}{Y}$$

where $Y = \hat{Y}_1 + \hat{Y}_2 + Y_b$.

In a similar way we can find the two remaining scattering parameters as

$$S_{12}^V(\gamma_1, \gamma_2) = \frac{2\hat{G}_2}{Y},$$

$$S_{22}^V(\gamma_1, \gamma_2) = \frac{2\hat{G}_2 - Y}{Y}.$$

From the relation between $\mathbf{S}(\gamma_1, \gamma_2)$ and $\mathbf{S}^V(\gamma_1, \gamma_2)$ parameters

$$S_{11}(\gamma_1,\gamma_2) = \frac{2\widehat{G}_1 - Y}{Y} \frac{\widehat{Y}_1}{\widehat{Y}_1^*}, \qquad S_{12}(\gamma_1,\gamma_2) = \frac{2\sqrt{(\widehat{G}_1\widehat{G}_2)}}{Y} \frac{\widehat{Y}_1\widehat{Y}_2}{|\widehat{Y}_1| \, |\widehat{Y}_2|},$$

$$S_{21}(\gamma_1,\gamma_2) = \frac{2\sqrt{(\widehat{G}_1\widehat{G}_2)}}{Y} \frac{\widehat{Y}_1\widehat{Y}_2}{|\widehat{Y}_1| \, |\widehat{Y}_2|}, \qquad S_{22}(\gamma_1,\gamma_2) = \frac{2\widehat{G}_2 - Y}{Y} \frac{\widehat{Y}_2}{\widehat{Y}_2^*}.$$

For the special case of $\widehat{Y}_1 = \widehat{Y}_2 = G_{01} = G_{02} = G_0 = 1/R_0$, we have

$$S_{11} = \frac{Y_b}{Y}, \qquad S_{12} = \frac{2G_0}{Y},$$

$$S_{21} = \frac{2G_0}{Y}, \qquad S_{22} = \frac{Y_b}{Y}$$

where $Y = 2G_0 + Y_b$, as in Example (2.1.3).

Example (6.1.4)

Calculate the scattering parameters of a shunt impedance $Z_b = 50 + j100$ for reference impedances (i) $\widehat{Z}_1 = \widehat{Z}_2 = 50$, (ii) $\widehat{Z}_1 = 50 + j25$ and $\widehat{Z}_2 = 50 - j25$, (iii) $\widehat{Z}_1 = 50 + j25$ and $\widehat{Z}_2 = 75 - j50$.

Solution

To solve this problem we directly substitute for the required parameters in the expressions found in our previous example.

(i) $Y = (0.004 - j0.008) + 0.020 + 0.020 = 0.044 - j0.008$, $\widehat{G}_1 = \widehat{G}_2 = 0.020$, hence after substitution

$$S_{11}(\gamma_1,\gamma_2) = S_{22}(\gamma_1,\gamma_2) = -0.12 + j0.16,$$

$$S_{12}(\gamma_1,\gamma_2) = S_{21}(\gamma_1,\gamma_2) = 0.88 + j0.16.$$

(ii) $Y = (0.004 - j0.008) + 0.016 - j0.008 + 0.016 + j0.008 = 0.036 - j0.008$, $G_1 = 0.016$, $G_2 = 0.016$, and after substitution

$$S_{11}(\gamma_1,\gamma_2) = 0.06 + j0.24, \qquad S_{12}(\gamma_1,\gamma_2) = 0.85 + j0.19,$$

$$S_{21}(\gamma_1,\gamma_2) = 0.85 + j0.19, \qquad S_{22}(\gamma_1,\gamma_2) = -0.24 - j0.01.$$

(iii) $Y = (0.004 - j0.008) + 0.016 - j0.008 + 0.009 + j0.006 = 0.029 - j0.010$, $G_1 = 0.016, G_2 = 0.009$, and after substitution

$$S_{11}(\gamma_1,\gamma_2) = 0.26 + j0.21, \qquad S_{12}(\gamma_1,\gamma_2) = 0.71 + j0.34,$$

$$S_{21}(\gamma_1,\gamma_2) = 0.71 + j0.34, \qquad S_{22}(\gamma_1,\gamma_2) = -0.34 - j0.33.$$

6.2 Derivation of the Scattering Parameters from the Impedance and Admittance Parameters

The generalized scattering parameters of a 2-port network are closely related to other network representations, such as the impedance and admittance parameters. In this section we derive the conversion relations for these different representations.

The conversion of the impedance parameters to the scattering parameters

From (6.1.15) and (6.1.16) and the definition of the impedance parameters

$$\mathbf{a}(\gamma_1,\gamma_2) = \frac{1}{2}\begin{bmatrix} 1/\sqrt{\hat{R}_1} & 0 \\ 0 & 1/\sqrt{\hat{R}_2} \end{bmatrix}\{\mathbf{Z} + \hat{\mathbf{Z}}\}\mathbf{I}, \tag{6.2.1}$$

$$\mathbf{b}(\gamma_1,\gamma_2) = \frac{1}{2}\begin{bmatrix} 1/\sqrt{\hat{R}_1} & 0 \\ 0 & 1/\sqrt{\hat{R}_2} \end{bmatrix}\{\mathbf{Z} - \hat{\mathbf{Z}}^*\}\mathbf{I} \tag{6.2.2}$$

and hence from the relation

$$\mathbf{b}(\gamma_1,\gamma_2) = \mathbf{S}(\gamma_1,\gamma_2)\mathbf{a}(\gamma_1,\gamma_2),$$

we have

$$S(\gamma_1,\gamma_2) = \begin{bmatrix} 1/\sqrt{\hat{R}_1} & 0 \\ 0 & 1/\sqrt{\hat{R}_2} \end{bmatrix} \left\{ Z - \hat{Z}^* \right\} \left\{ Z + \hat{Z} \right\}^{-1} \begin{bmatrix} \sqrt{\hat{R}_1} & 0 \\ 0 & \sqrt{\hat{R}_2} \end{bmatrix} \quad (6.2.3)$$

relating the scattering parameters to the impedance parameters as

$$S(\gamma_1,\gamma_2) = \frac{1}{\Sigma_{ZS}} \begin{bmatrix} \dfrac{Z_{11} - \hat{Z}_1^*}{\sqrt{\hat{R}_1}} & \dfrac{Z_{12}}{\sqrt{\hat{R}_1}} \\ \dfrac{Z_{21}}{\sqrt{\hat{R}_2}} & \dfrac{Z_{22} - \hat{Z}_2^*}{\sqrt{\hat{R}_2}} \end{bmatrix} \begin{bmatrix} \sqrt{\hat{R}_1}(Z_{22} + Z_2) & -\sqrt{\hat{R}_2}Z_{12} \\ -\sqrt{\hat{R}_2}Z_{21} & \sqrt{\hat{R}_2}(Z_{11} + Z_1) \end{bmatrix}$$

where

$$\Sigma_{ZS} = \hat{Z}_1\hat{Z}_2 + \hat{Z}_2 Z_{11} + \hat{Z}_1 Z_{22} + \Delta_Z \quad (6.2.4)$$

and

$$\Delta_Z = Z_{11}Z_{22} - Z_{12}Z_{21}.$$

After matrix multiplication, the scattering parameters are found in terms of the impedance parameters as

$$S_{11}(\gamma_1,\gamma_2) = \frac{1}{\Sigma_{ZS}}(\hat{Z}_2 Z_{11} - \hat{Z}_1^* Z_{22} - \hat{Z}_1^* \hat{Z}_2 + \Delta_Z), \quad (6.2.5a)$$

$$S_{12}(\gamma_1,\gamma_2) = \frac{2}{\Sigma_{ZS}}\sqrt{(\hat{R}_1\hat{R}_2)}Z_{12}, \quad (6.2.5b)$$

$$S_{21}(\gamma_1,\gamma_2) = \frac{2}{\Sigma_{ZS}}\sqrt{(\hat{R}_2\hat{R}_1)}Z_{21}, \quad (6.2.5c)$$

$$S_{22}(\gamma_1,\gamma_2) = \frac{1}{\Sigma_{ZS}}(\hat{Z}_1 Z_{22} - \hat{Z}_2^* Z_{11} - \hat{Z}_2^* \hat{Z}_1 + \Delta_Z). \quad (6.2.5d)$$

The conversion of the scattering parameters to the impedance parameters

To find the impedance parameters in terms of the scattering parameters, (6.2.3) can be written as

$$\begin{bmatrix} 1/\sqrt{\hat{R}_1} & 0 \\ 0 & 1/\sqrt{\hat{R}_2} \end{bmatrix} \{ \mathbf{Z} - \hat{\mathbf{Z}}^* \} = \mathbf{S}(\gamma_1, \gamma_2) \begin{bmatrix} 1/\sqrt{\hat{R}_1} & 0 \\ 0 & 1/\sqrt{\hat{R}_2} \end{bmatrix} \{ \mathbf{Z} + \hat{\mathbf{Z}} \}.$$

With some rearrangement we can find $\mathbf{Z}$ as

$$\mathbf{Z} = \begin{bmatrix} \sqrt{\hat{R}_1} & 0 \\ 0 & \sqrt{\hat{R}_2} \end{bmatrix} \{ \mathbf{U} - \mathbf{S}(\gamma_1, \gamma_2) \}^{-1} \times$$

$$\left\{ \begin{bmatrix} 1/\sqrt{\hat{R}_1} & 0 \\ 0 & 1/\sqrt{\hat{R}_2} \end{bmatrix} \hat{\mathbf{Z}}^* + \mathbf{S}(\gamma_1, \gamma_2) \begin{bmatrix} 1/\sqrt{\hat{R}_1} & 0 \\ 0 & 1/\sqrt{\hat{R}_2} \end{bmatrix} \hat{\mathbf{Z}} \right\}.$$

After matrix inversion and multiplication we find the impedance parameters in terms of the scattering parameters as

$$Z_{11} = \frac{\hat{Z}_1}{\Sigma_{SZ}} [\zeta(\gamma_1) + S_{11}(\gamma_1, \gamma_2) - \zeta(\gamma_1) S_{22}(\gamma_1, \gamma_2) - \Delta_S(\gamma_1, \gamma_2)], \quad (6.2.6a)$$

$$Z_{12} = \frac{2}{\Sigma_{SZ}} \sqrt{(\hat{R}_1 \hat{R}_2)} S_{12}(\gamma_1, \gamma_2), \quad (6.2.6b)$$

$$Z_{21} = \frac{2}{\Sigma_{SZ}} \sqrt{(\hat{R}_2 \hat{R}_1)} S_{21}(\gamma_1, \gamma_2), \quad (6.2.6c)$$

$$Z_{22} = \frac{\hat{Z}_2}{\Sigma_{SZ}} [\zeta(\gamma_2) + S_{22}(\gamma_1, \gamma_2) - \zeta(\gamma_2) S_{11}(\gamma_1, \gamma_2) - \Delta_S(\gamma_1, \gamma_2)] \quad (6.2.6d)$$

where

$$\Sigma_{SZ}(\gamma_1,\gamma_2) = 1 - S_{11}(\gamma_1,\gamma_2) - S_{22}(\gamma_1,\gamma_2) + \Delta_S(\gamma_1,\gamma_2) \qquad (6.2.7)$$

and

$$\Delta_S(\gamma_1,\gamma_2) = S_{11}(\gamma_1,\gamma_2)S_{22}(\gamma_1,\gamma_2) - S_{12}(\gamma_1,\gamma_2)S_{21}(\gamma_1,\gamma_2).$$

In (6.2.6), $\zeta(\gamma_1) = (\hat{Z}_1^* / \hat{Z}_1)$ and $\zeta(\gamma_2) = (\hat{Z}_2^* / \hat{Z}_2)$.

The conversion of the Y-parameters to the scattering parameters

From (6.1.15) and (6.1.16) and the definition of the admittance parameters

$$\mathbf{a}(\gamma_1,\gamma_2) = \frac{1}{2}\begin{bmatrix} 1/\sqrt{\hat{R}_1} & 0 \\ 0 & 1/\sqrt{\hat{R}_2} \end{bmatrix} \hat{\mathbf{Z}}\{\hat{\mathbf{Y}} + \mathbf{Y}\}\mathbf{V} \qquad (6.2.8)$$

and

$$\mathbf{b}(\gamma_1,\gamma_2) = \frac{1}{2}\begin{bmatrix} 1/\sqrt{\hat{R}_1} & 0 \\ 0 & 1/\sqrt{\hat{R}_2} \end{bmatrix} \hat{\mathbf{Z}}^*\{\hat{\mathbf{Y}}^* - \mathbf{Y}\}\mathbf{V} \qquad (6.2.9)$$

where

$$\hat{\mathbf{Y}} = \hat{\mathbf{Z}}^{-1} = \begin{bmatrix} \hat{Y}_1 & 0 \\ 0 & \hat{Y}_2 \end{bmatrix}.$$

Hence from

$$\mathbf{b}(\gamma_1,\gamma_2) = \mathbf{S}(\gamma_1,\gamma_2)\mathbf{a}(\gamma_1,\gamma_2),$$

we have

$$\mathbf{S}(\gamma_1,\gamma_2) = \begin{bmatrix} \hat{Z}_1^*/\sqrt{\hat{R}_1} & 0 \\ 0 & \hat{Z}_2^*/\sqrt{\hat{R}_2} \end{bmatrix} \{\hat{\mathbf{Y}}^* - \mathbf{Y}\}\{\hat{\mathbf{Y}} + \mathbf{Y}\}^{-1} \begin{bmatrix} \sqrt{\hat{R}_1}\hat{Y}_1 & 0 \\ 0 & \sqrt{\hat{R}_2}\hat{Y}_2 \end{bmatrix} \qquad (6.2.10)$$

or

$$S(\gamma_1, \gamma_2) =$$

$$\frac{1}{\Sigma_{YS}} \begin{bmatrix} \dfrac{\hat{Z}_1^*(\hat{Y}_1^* - Y_{11})}{\sqrt{\hat{R}_1}} & \dfrac{-\hat{Z}_1^* Y_{12}}{\sqrt{\hat{R}_1}} \\[4mm] \dfrac{-\hat{Z}_2^* Y_{21}}{\sqrt{\hat{R}_2}} & \dfrac{\hat{Z}_2^*(\hat{Y}_2^* - Y_{22})}{\sqrt{\hat{R}_2}} \end{bmatrix} \begin{bmatrix} \sqrt{\hat{R}_1}\,\hat{Y}_1(Y_{22} + \hat{Y}_2) & -\sqrt{\hat{R}_2}\,\hat{Y}_2 Y_{12} \\[4mm] -\sqrt{\hat{R}_1}\,\hat{Y}_1 Y_{21} & \sqrt{\hat{R}_2}\,\hat{Y}_2(Y_{11} + \hat{Y}_1) \end{bmatrix}$$

$$(6.2.11)$$

where

$$\Sigma_{YS} = \hat{Y}_1 \hat{Y}_2 + \hat{Y}_2 Y_{11} + \hat{Y}_1 Y_{22} + \Delta_Y$$

and

$$\Delta_Y = Y_{11} Y_{22} - Y_{12} Y_{21}.$$

Hence the scattering parameters in terms of the admittance parameters are given by the following expressions:

$$S_{11}(\gamma_1, \gamma_2) = \frac{\zeta(\gamma_1)}{\Sigma_{YS}} (\hat{Y}_2 \hat{Y}_1^* - Y_{11} \hat{Y}_2 + \hat{Y}_1^* Y_{22} - \Delta_Y), \qquad (6.2.12a)$$

$$S_{12}(\gamma_1, \gamma_2) = -\frac{2}{\Sigma_{YS}} \frac{\hat{Y}_1}{|\hat{Y}_1|} \frac{\hat{Y}_2}{|\hat{Y}_2|} \sqrt{(\hat{G}_1 \hat{G}_2)}\, Y_{12}, \qquad (6.2.12b)$$

$$S_{21}(\gamma_1, \gamma_2) = -\frac{2}{\Sigma_{YS}} \frac{\hat{Y}_1}{|\hat{Y}_1|} \frac{\hat{Y}_2}{|\hat{Y}_2|} \sqrt{(\hat{G}_1 \hat{G}_2)}\, Y_{21}, \qquad (6.2.12c)$$

$$S_{22}(\gamma_1, \gamma_2) = \frac{\zeta(\gamma_2)}{\Sigma_{YS}} (\hat{Y}_1 \hat{Y}_2^* - Y_{22} \hat{Y}_1 + \hat{Y}_2^* Y_{11} - \Delta_Y) \qquad (6.2.12d)$$

where $\hat{G}_1 = \mathrm{Re}(\hat{Y}_1)$, $\hat{G}_2 = \mathrm{Re}(\hat{Y}_2)$, $\zeta(\gamma_1) = \hat{Y}_1 / \hat{Y}_1^*$ and $\zeta(\gamma_2) = \hat{Y}_2 / \hat{Y}_2^*$.

The conversion of the scattering parameters to the admittance parameters

To find the admittance parameters in terms of the scattering parameters, we write (6.2.10) as

$$
\begin{bmatrix} \hat{Z}_1^*/\sqrt{\hat{R}_1} & 0 \\ 0 & \hat{Z}_2^*/\sqrt{\hat{R}_2} \end{bmatrix} \{\hat{\mathbf{Y}}^* - \mathbf{Y}\} = \mathbf{S}(\gamma_1, \gamma_2) \begin{bmatrix} 1/\sqrt{\hat{R}_1}\,\hat{Y}_1 & 0 \\ 0 & 1/\sqrt{\hat{R}_2}\,\hat{Y}_2 \end{bmatrix} \{\hat{\mathbf{Y}} + \mathbf{Y}\}.
$$

Hence after rearrangement we can find **Y** as

$$
\mathbf{Y} = \left\{ \begin{bmatrix} \hat{Z}_1^*/\sqrt{\hat{R}_1} & 0 \\ 0 & \hat{Z}_2^*/\sqrt{\hat{R}_2} \end{bmatrix} \mathbf{U} + \mathbf{S}(\gamma_1, \gamma_2) \begin{bmatrix} 1/\sqrt{\hat{R}_1}\,\hat{Y}_1 & 0 \\ 0 & 1/\sqrt{\hat{R}_2}\,\hat{Y}_2 \end{bmatrix} \right\}^{-1} \times
$$

$$
\left\{ \begin{bmatrix} \hat{Z}_1^*/\sqrt{\hat{R}_1} & 0 \\ 0 & \hat{Z}_2^*/\sqrt{R_2} \end{bmatrix} \hat{\mathbf{Y}}^* - \mathbf{S}(\gamma_1, \gamma_2) \begin{bmatrix} 1/\sqrt{\hat{R}_1}\,Y_1 & 0 \\ 0 & 1/\sqrt{\hat{R}_2}\,\hat{Y}_2 \end{bmatrix} \hat{\mathbf{Y}} \right\}.
$$

By matrix inversion and multiplication we find the admittance parameters in terms of the scattering parameters as

$$
Y_{11} = \frac{\hat{Y}_1}{\Sigma_{SY}} [\zeta(\gamma_2) + S_{22}(\gamma_1, \gamma_2) - \zeta(\gamma_2) S_{11}(\gamma_1, \gamma_2) - \Delta_S(\gamma_1, \gamma_2)], \quad (6.2.13a)
$$

$$
Y_{12} = -\frac{2}{\Sigma_{SY}} \frac{\hat{Y}_1 \hat{Y}_2}{|\hat{Y}_1||\hat{Y}_2|} \sqrt{(\hat{G}_1 \hat{G}_2)} S_{12}(\gamma_1, \gamma_2), \quad (6.2.13b)
$$

$$
Y_{21} = -\frac{2}{\Sigma_{SY}} \frac{\hat{Y}_1 \hat{Y}_2}{|\hat{Y}_1||\hat{Y}_2|} \sqrt{(\hat{G}_1 \hat{G}_2)} S_{21}(\gamma_1, \gamma_2), \quad (6.2.13c)
$$

$$
Y_{22} = \frac{\hat{Y}_2}{\Sigma_{SY}} [\zeta(\gamma_1) + S_{11}(\gamma_1, \gamma_2) - \zeta(\gamma_1) S_{22}(\gamma_1, \gamma_2) - \Delta_S(\gamma_1, \gamma_2)] \quad (6.2.13d)
$$

where

$$\Sigma_{SY}(\gamma_1,\gamma_2) = \zeta(\gamma_1)\zeta(\gamma_2) + \zeta(\gamma_2)S_{11}(\gamma_1,\gamma_2)$$

$$+ \zeta(\gamma_1)S_{22}(\gamma_1,\gamma_2) + \Delta_S(\gamma_1,\gamma_2). \qquad (6.2.14)$$

Example (6.2.1)

Using the conversion between the admittance parameters and the scattering parameters, find the scattering parameters of Example (6.1.2).

Solution

From the definition of the admittance parameters, the admittance parameters of a series impedance Z_a are given by the expressions

$$Y_{11} = \frac{1}{Z_a}, \qquad Y_{12} = -\frac{1}{Z_a},$$

$$Y_{21} = -\frac{1}{Z_a}, \qquad Y_{22} = \frac{1}{Z_a}.$$

With $Z_a = 100 + j50$, $Y_a = 0.008 - j0.004$ and we have

(i) $\hat{Z}_1 = \hat{Z}_2 = 50$, $\hat{Y}_1 = \hat{Y}_2 = 0.02$, $\Sigma_{YS} = 0.00072 - j0.00016$, and hence from (6.2.12)

$$S_{11} = S_{22} = 0.53 + j0.12,$$

$$S_{12} = S_{21} = 0.47 - j0.12.$$

(ii) $\hat{Z}_1 = 50 + j25$, $\hat{Z}_2 = 50 - j25$, $\hat{Y}_1 = 0.016 - j0.008$, $\hat{Y}_2 = 0.016 + j0.008$, $\Sigma_{YS} = 0.000576 - j0.000128$, and again from (6.2.12)

$$S_{11}(\gamma_1,\gamma_2) = S_{22}(\gamma_1,\gamma_2) = 0.53 + j0.12,$$

$$S_{12}(\gamma_1,\gamma_2) = S_{21}(\gamma_1,\gamma_2) = 0.47 - j0.12.$$

(iii) $\hat{Z}_1 = 50 + j25$, $\hat{Z}_2 = 75 - j50$, $\hat{Y}_1 = 0.016 - j0.008$, $\hat{Y}_2 = 0.00923 + j0.00615$
$\Sigma_{YS} = 0.000391 - j0.000091$, and from (6.2.12)

$$S_{11}(\gamma_1, \gamma_2) = 0.56 + j0.05, \qquad S_{22}(\gamma_1, \gamma_2) = 0.54 - j0.06,$$

$$S_{12}(\gamma_1, \gamma_2) = 0.54 - j0.06, \qquad S_{21}(\gamma_1, \gamma_2) = 0.34 + j0.07.$$

The above parameters are identical to the scattering parameters found in Example (6.1.2).

Example (6.2.2)

Using the conversion between the impedance parameters and the scattering matrix parameters, find the scattering matrix parameters of Example (6.1.4).

Solution

From the definition of impedance parameter and the impedance parameters of a shunt impedance Z_b are given by

$$Z_{11} = Z_{12} = Z_{21} + Z_{22} = Z_b.$$

With $Z_b = 50 + j100$ we have

(i) $\hat{Z}_1 = \hat{Z}_2 = 50$, $\Sigma_{ZS} = 7500 + j10000$, and hence from (6.2.5)

$$S_{11} = S_{22} = -0.12 + j0.16,$$

$$S_{12} = S_{21} = 0.88 + j0.16.$$

(ii) $\hat{Z}_1 = 50 + j25$, $\hat{Z}_2 = 50 - j25$, $\Sigma_{ZS} = 8125 + j10000$, hence from (6.2.5)

$$S_{11}(\gamma_1, \gamma_2) = 0.06 + j0.24, \qquad S_{12}(\gamma_1, \gamma_2) = 0.85 + j0.19,$$

$$S_{21}(\gamma_1, \gamma_2) = 0.85 + j0.19, \qquad S_{22}(\gamma_1, \gamma_2) = -0.24 - j0.01.$$

(iii) $\hat{Z}_1 = 50 + j25$, $\hat{Z}_2 = 75 - j50$, $\Sigma_{ZS} = 13750 + j10625$, hence from (6.2.5)

$$S_{11}(\gamma_1, \gamma_2) = 0.26 + j0.21, \qquad S_{12}(\gamma_1, \gamma_2) = 0.71 + j0.34,$$

$$S_{21}(\gamma_1, \gamma_2) = 0.71 + j0.34, \qquad S_{22}(\gamma_1, \gamma_2) = -0.34 - j0.33.$$

The above parameters are identical to the scattering parameters found in Example (6.1.4).

Example (6.2.3)

Find the scattering parameters for the π-section of Example (2.3.4) for (i) $\hat{Z}_1 = \hat{Z}_2 = 50$, (ii) $\hat{Z}_1 = 50 + j25$ and $\hat{Z}_2 = 50 - j25$, (iii) $\hat{Z}_1 = 50 + j25$, $\hat{Z}_2 = 75 - j50$.

Solution

(i) Substituting for values of Z_{11}, Z_{12}, Z_{21}, Z_{22} into (6.2.5) and with $\Sigma_{ZS} = 10956 + j7426$, we find

$$S_{11} = 0.17 + j0.40, \qquad S_{12} = 0.290 - j0.01,$$

$$S_{21} = 0.29 - j0.01, \qquad S_{22} = 0.104 + j0.84.$$

(ii)   In this case $\Sigma_{ZS} = 12574 + j7831$ and again substituting into (6.2.5)

$$S_{11}(\gamma_1, \gamma_2) = 0.34 + j0.47, \qquad S_{12}(\gamma_1, \gamma_2) = 026 + j0.00,$$

$$S_{21}(\gamma_1, \gamma_2) = 0.26 + j0.00, \qquad S_{22}(\gamma_1, \gamma_2) = 0.11 - j0.10.$$

(iii) With $\Sigma_{ZS} = 17243 + j7279$ we have

$$S_{11}(\gamma_1, \gamma_2) = 0.35 + j0.46, \qquad S_{12}(\gamma_1, \gamma_2) = 0.25 + 0.04,$$

$$S_{21}(\gamma_1, \gamma_2) = 0.25 + j0.09, \qquad S_{22}(\gamma_1, \gamma_2) = -0.03 - j0.28.$$

6.3 Conversion of the Scattering Matrix in a System of Resistive References to other Systems of Generalized Reference Impedances

In this section appropriate expressions are given for the conversion of the scattering parameters in a system of standard resistive references, to the scattering parameters in a generalized system. The measurements of the scattering parameters are made with the network connected via certain guiding structures to generators or load terminations. The characteristic impedance of the waveguides and the terminating impedances are made equal for each port, and are assumed to be resistive. Hence the expressions that will be given would convert the measured scattering parameters with resistive references to the scattering parameters in any other system of reference impedances.

As before, we denote the standard resistive references as R_{01} and R_{02} for port 1 and port 2 respectively. When $\hat{Z}_1 = R_{01}$, $\gamma_1 = 0$, and we denote the scattering matrix by $\mathbf{S}(0,\gamma_2)$. Similarly when $\hat{Z}_2 = R_{02}$, $\gamma_2 = 0$ and the scattering matrix can be denoted by $\mathbf{S}(\gamma_1,0)$. When both γ_1 and γ_2 are zero, however, we denote the scattering matrix by $\mathbf{S}$ instead of $\mathbf{S}(0,0)$ for convenience. In this section, in addition to the conversion expressions of $\mathbf{S}$ to $\mathbf{S}(\gamma_1,\gamma_2)$, we also give the useful conversion expressions of $\mathbf{S}(\gamma_1,0)$ to $\mathbf{S}(\gamma_1,\gamma_2)$, $\mathbf{S}$ to $\mathbf{S}(\gamma_1,0)$, $\mathbf{S}(0,\gamma_2)$ to $\mathbf{S}(\gamma_1,\gamma_2)$ and $\mathbf{S}$ to $\mathbf{S}(0,\gamma_2)$.

For the conversion of the scattering parameters we make use of the impedance parameters, which unlike the scattering parameters, are uniquely defined, independent of any reference impedance. From (6.2.6) we write the impedance parameters in terms of the scattering parameters in a system of standard resistive references as

$$Z_{11} = \frac{R_{01}}{\Sigma_{SZ}}(1 - S_{22} + S_{11} - \Delta_S), \tag{6.3.1a}$$

$$Z_{12} = \frac{2\sqrt{(R_{01}R_{02})}}{\Sigma_{SZ}} S_{12}, \tag{6.3.1b}$$

$$Z_{21} = \frac{2\sqrt{(R_{01}R_{02})}}{\Sigma_{SZ}} S_{21}, \tag{6.3.1c}$$

$$Z_{22} = \frac{R_{02}}{\Sigma_{SZ}}(1 - S_{11} + S_{22} - \Delta_S) \tag{6.3.1d}$$

where

$$\Sigma_{SZ} = 1 - S_{11} - S_{22} + \Delta_S.$$

In a general system of reference impedances, the scattering parameters in terms of the impedance parameters were given by (6.2.5) as

$$S_{11}(\gamma_1,\gamma_2) = \frac{1}{\Sigma_{ZS}}(\hat{Z}_2 Z_{11} - \hat{Z}_1^* Z_{22} - \hat{Z}_1^* \hat{Z}_2 + \Delta_Z), \tag{6.3.2a}$$

$$S_{12}(\gamma_1,\gamma_2) = \frac{2}{\Sigma_{ZS}}\sqrt{(\hat{R}_1\hat{R}_2)}Z_{12}, \tag{6.3.2b}$$

$$S_{21}(\gamma_1, \gamma_2) = \frac{2}{\Sigma_{ZS}} \sqrt{(\hat{R}_1 \hat{R}_2)} Z_{21}, \tag{6.3.2c}$$

$$S_{22}(\gamma_1, \gamma_2) = \frac{1}{\Sigma_{ZS}} (\hat{Z}_1 Z_{22} - \hat{Z}_2^* Z_{11} - \hat{Z}_2^* \hat{Z}_1 + \Delta_Z) \tag{6.3.2d}$$

where Σ_{ZS} is defined by (6.2.4).

Substituting for Z_{11}, Z_{12}, Z_{21} and Z_{22} from (6.3.1) to (6.3.2) we find $S(\gamma_1, \gamma_2)$ parameters in terms of the measured scattering parameters as

$$S_{11}(\gamma_1, \gamma_2) = \eta_1 \frac{(1 - \gamma_2 S_{22})(S_{11} - \gamma_1^*) + \gamma_2 S_{12} S_{21}}{(1 - \gamma_2 S_{22})(1 - \gamma_1 S_{11}) - \gamma_1 \gamma_2 S_{12} S_{21}}, \tag{6.3.3a}$$

$$S_{12}(\gamma_1, \gamma_2) = \Lambda_1^* \Lambda_2^* \frac{S_{12}}{(1 - \gamma_2 S_{22})(1 - \gamma_1 S_{11}) - \gamma_1 \gamma_2 S_{12} S_{21}}, \tag{6.3.3b}$$

$$S_{21}(\gamma_1, \gamma_2) = \Lambda_1^* \Lambda_2^* \frac{S_{21}}{(1 - \gamma_2 S_{22})(1 - \gamma_1 S_{11}) - \gamma_1 \gamma_2 S_{12} S_{21}}, \tag{6.3.3c}$$

$$S_{22}(\gamma_1, \gamma_2) = \eta_2 \frac{(1 - \gamma_1 S_{11})(S_{22} - \gamma_2^*) + \gamma_1 S_{12} S_{21}}{(1 - \gamma_2 S_{22})(1 - \gamma_1 S_{11}) - \gamma_1 \gamma_2 S_{12} S_{21}} \tag{6.3.3d}$$

where

$$\Lambda_1 = (1 - \gamma_1^*) \left[\frac{1 - \gamma_1 \gamma_1^*}{(1 - \gamma_1)(1 - \gamma_1^*)} \right]^{1/2} \tag{6.3.4a}$$

and

$$\Lambda_2 = (1 - \gamma_2^*) \left[\frac{1 - \gamma_2 \gamma_2^*}{(1 - \gamma_2)(1 - \gamma_2^*)} \right]^{1/2} \tag{6.3.4b}$$

and also $\eta_1 = \Lambda_1^* / \Lambda_1$ and $\eta_2 = \Lambda_2^* / \Lambda_2$.

We note that

$$|\Lambda_1| = |\Lambda_1^*| = \sqrt{(1 - |\gamma_1|^2)} \quad \text{and} \quad |\Lambda_2| = |\Lambda_2^*| = \sqrt{(1 - |\gamma_2|^2)}.$$

Similarly, by using the impedance parameters, we can convert $\mathbf{S}(\gamma_1,0)$ parameters to $\mathbf{S}(\gamma_1,\gamma_2)$ parameters as given by the following expressions

$$S_{11}(\gamma_1,\gamma_2) = \frac{S_{11}(\gamma_1,0) - \gamma_2 \Delta(\gamma_1,0)}{1 - \gamma_2 S_{22}(\gamma_1,0)}, \tag{6.3.5a}$$

$$S_{12}(\gamma_1,\gamma_2) = \Lambda_2^* \frac{S_{12}(\gamma_1,0)}{1 - \gamma_2 S_{22}(\gamma_1,0)}, \tag{6.3.5b}$$

$$S_{21}(\gamma_1,\gamma_2) = \Lambda_2^* \frac{S_{21}(\gamma_1,0)}{1 - \gamma_2 S_{22}(\gamma_1,0)}, \tag{6.3.5c}$$

$$S_{22}(\gamma_1,\gamma_2) = \eta_2 \frac{S_{22}(\gamma_1,0) - \gamma_2^*}{1 - \gamma_2 S_{22}(\gamma_1,0)} \tag{6.3.5d}$$

where

$$\Delta_S(\gamma_1,0) = S_{11}(\gamma_1,0)S_{22}(\gamma_1,0) - S_{12}(\gamma_1,0)S_{21}(\gamma_1,0).$$

Alternatively, by again using the impedance parameters, we can convert $\mathbf{S}(0,\gamma_2)$ parameters to $\mathbf{S}(\gamma_1,\gamma_2)$ parameters in the form of

$$S_{11}(\gamma_1,\gamma_2) = \eta_1 \frac{S_{11}(0,\gamma_2) - \gamma_1^*}{1 - \gamma_1 S_{11}(0,\gamma_2)}, \tag{6.3.6a}$$

$$S_{12}(\gamma_1,\gamma_2) = \Lambda_1^* \frac{S_{12}(0,\gamma_2)}{1 - \gamma_1 S_{11}(0,\gamma_2)}, \tag{6.3.6b}$$

$$S_{21}(\gamma_1,\gamma_2) = \Lambda_1^* \frac{S_{21}(0,\gamma_2)}{1 - \gamma_1 S_{11}(0,\gamma_2)}, \tag{6.3.6c}$$

$$S_{22}(\gamma_1,\gamma_2) = \frac{S_{22}(0,\lambda_2) - \gamma_1 \Delta_S(0,\gamma_2)}{1 - \gamma_1 S_{11}(0,\gamma_2)} \tag{6.3.6d}$$

where

$$\Delta_S(0,\gamma_2) = S_{11}(0,\gamma_2)S_{22}(0,\gamma_2) - S_{12}(0,\gamma_2)S_{21}(0,\gamma_2).$$

In turn, the elements of $S(\gamma_1, 0)$ can be written in terms of the elements of S as

$$S_{11}(\gamma_1, 0) = \eta_1 \frac{S_{11} - \gamma_1^*}{(1 - \gamma_1 S_{11})}, \tag{6.3.7a}$$

$$S_{12}(\gamma_1, 0) = \Lambda_1^* \frac{S_{12}}{(1 - \gamma_1 S_{11})}, \tag{6.3.7b}$$

$$S_{21}(\gamma_1, 0) = \Lambda_1^* \frac{S_{21}}{(1 - \gamma_1 S_{11})}, \tag{6.3.7c}$$

$$S_{22}(\gamma_1, 0) = \frac{S_{22} - \gamma_1 \Delta_S}{(1 - \gamma_1 S_{11})} \tag{6.3.7d}$$

where

$$\Delta_S = (S_{11}S_{22} - S_{12}S_{21}).$$

Similarly $S(0, \gamma_2)$ can be written in terms of the measured scattering parameters as

$$S_{11}(0, \gamma_2) = \frac{S_{11} - \gamma_2 \Delta_S}{(1 - \gamma_2 S_{22})}, \tag{6.3.8a}$$

$$S_{12}(0, \gamma_2) = \Lambda_2^* \frac{S_{12}}{(1 - \gamma_2 S_{22})}, \tag{6.3.8b}$$

$$S_{21}(0, \gamma_2) = \Lambda_2^* \frac{S_{21}}{(1 - \gamma_2 S_{22})}, \tag{6.3.8c}$$

$$S_{22}(0, \gamma_2) = \eta_2 \frac{S_{22} - \gamma_2^*}{(1 - \gamma_2 S_{22})}. \tag{6.3.8d}$$

Example (6.3.1)

In Example (6.1.2), the scattering parameters of a series impedance $Z_a = 100 + j50$ for reference impedances (i) $\hat{Z}_1 = \hat{Z}_2 = 50$ and (ii) $\hat{Z}_1 = 50 + j25$, $\hat{Z}_2 = 75 - j50$ were given as

(i) $S_{11} = S_{22} = 0.53 + j0.12,$ $S_{12} = S_{21} = 0.47 - j0.12$

and

(ii) $S_{11}(\gamma_1, \gamma_2) = 0.56 + j0.05,$ $S_{12}(\gamma_1, \gamma_2) = 0.54 - j0.06,$

$S_{21}(\gamma_1, \gamma_2) = 0.54 - j0.06,$ $S_{22}(\gamma_1, \gamma_2) = 0.34 + j0.07,$

convert (i), using the conversion expressions (6.3.8) and (6.3.6), and verify (ii).

Solution

We first convert the scattering parameters in a system of reference $\hat{Z}_1 = \hat{Z}_2 = 50$ to scattering parameters in a system of reference $\hat{Z}_1 = 50,$ $\hat{Z}_2 = 75 - j50$, using (6.3.8). The required parameters are $\gamma_2 = 0.415\angle -41.6°,$ $\Lambda_2 = 0.910\angle -21.8°$ and $1 - \gamma_2 S_{22}$ $= 0.811\angle 7.7°$. Having found the elements of $\mathbf{S}(0, \gamma_2)$, (6.3.6) gives $\mathbf{S}(\gamma_1, \gamma_2)$ parameters as in (ii).

Example **(6.3.2)**

In Example (6.1.4) the scattering parameters of a shunt impedance $Z_b = 50 + j100$ for reference impedances (i) $\hat{Z}_1 = \hat{Z}_2 = 50$, (ii) $\hat{Z}_1 = 50 + j25$ and $\hat{Z}_2 = 75 - j50$ were given respectively as

(i) $S_{11} = S_{22} = -0.12 + j0.16,$ $S_{12} = S_{21} = 0.88 + j0.16$

and

(ii) $S_{11}(\gamma_1, \gamma_2) = 0.26 + j0.21,$ $S_{12}(\gamma_1, \gamma_2) = 0.71 + j0.34,$

$S_{21}(\gamma_1, \gamma_2) = 0.71 + j0.34,$ $S_{22}(\gamma_1, \gamma_2) = -0.34 - j0.33,$

convert (i), using conversion expressions (6.3.3) and verify (ii).

Solution

The required parameters for this conversion are $\gamma_1 = 0.242\angle 76.0°,$ $\Lambda_1 = 0.970\angle 14.0°,$ $\gamma_2 = 0.415\angle -41.6°,$ $\Lambda_2 = 0.910\angle -21.8°$ and $(1 - S_{11}\gamma_1)(1 - \gamma_2 S_{22}) - \gamma_1 \gamma_2 S_{12} S_{21} =$ $1.02\angle -3.5°$. Using the above parameters, the scattering parameters for case (ii) are readily verified.

Example (6.3.3)

The scattering parameters of a network in a system of resistive reference impedances $\hat{Z}_1 = \hat{Z}_2 = 50$ is given as

$$S_{11} = 0.80\angle 30°, \qquad S_{12} = 0.05\angle -60°,$$

$$S_{21} = 2.00\angle 0°, \qquad S_{22} = 0.60\angle 45°.$$

Find the scattering parameters in systems of reference impedances $(\hat{Z}_1 = 100 - j50,$ $\hat{Z}_2 = 50)$, $(\hat{Z}_1 = 50, \quad \hat{Z}_2 = 75 + j25)$ and $(\hat{Z}_1 = 100 - j50, \quad \hat{Z}_2 = 75 + j25)$.

Solution

For the given reference impedances we have the reference reflection coefficients as ($\gamma_1 =$ $0.447 \angle -26.6°, \quad \gamma_2 = 0$), ($\gamma_1 = 0, \quad \gamma_2 = 0.277\angle 33.7°$) and ($\gamma_1 = 0.447\angle -26.6°$, $\gamma_2 = 0.277 \angle 33.7°$) respectively.

By direct substitution we can find the new scattering parameters as given below.

(γ_1, γ_2)	$(0,0)$	$(\gamma_1, 0)$	$(0, \gamma_2)$	(γ_1, γ_2)
$S_{11}(\gamma_1, \gamma_2)$	$0.80 \angle 30.0°$	$0.551\angle 73.1°$	$0.820 \angle 28.6°$	$0.589 \angle 69.0°$
$S_{12}(\gamma_1, \gamma_2)$	$0.05 \angle -60°$	$0.070\angle -39.7°$	$0.049 \angle -61.7°$	$0.069 \angle -42.1°$
$S_{21}(\gamma_1, \gamma_2)$	$2.00 \angle 0.0°$	$2.781 \angle 20.3°$	$1.959 \angle -1.7°$	$2.764 \angle 17.9°$
$S_{22}(\gamma_1, \gamma_2)$	$0.60 \angle 45.0°$	$0.558 \angle 39.5°$	$0.621 \angle 58.4°$	$0.566 \angle 54.7°$

6.4 Special Parameters

In the previous chapters we considered a number of parameters that had significance in the discussions of simultaneous conjugate matching and stability. These parameters were denoted as B_L, B_S, C_L, C_S, Δ_S and K. We now formally extend the definitions of these parameters and express them in terms of the generalized scattering parameters. To this end we define the following generalized parameters:

$$B_L(\gamma_1, \gamma_2) = \frac{1 - |S_{11}(\gamma_1, \gamma_2)|^2 + |S_{22}(\gamma_1, \gamma_2)|^2 - |\Delta_S(\gamma_1, \gamma_2)|^2}{2|S_{12}(\gamma_1, \gamma_2)S_{21}(\gamma_1, \gamma_2)|},$$

$$B_S(\gamma_1, \gamma_2) = \frac{1 - |S_{22}(\gamma_1, \gamma_2)|^2 + |S_{11}(\gamma_1, \gamma_2)|^2 - |\Delta_S(\gamma_1, \gamma_2)|^2}{2|S_{12}(\gamma_1, \gamma_2)S_{21}(\gamma_1, \gamma_2)|},$$

$$C_L(\gamma_1, \gamma_2) = \frac{S_{22}(\gamma_1, \gamma_2) - S_{11}(\gamma_1, \gamma_2)^* \Delta_S(\gamma_1, \gamma_2)}{|S_{12}(\gamma_1, \gamma_2)S_{21}(\gamma_1, \gamma_2)|},$$

$$C_S(\gamma_1, \gamma_2) = \frac{S_{11}(\gamma_1, \gamma_2) - S_{22}(\gamma_1, \gamma_2)^* \Delta_S(\gamma_1, \gamma_2)}{|S_{12}(\gamma_1, \gamma_2)S_{21}(\gamma_1, \gamma_2)|},$$

$$\Delta_S(\gamma_1, \gamma_2) = S_{11}(\gamma_1, \gamma_2)S_{22}(\gamma_1, \gamma_2) - S_{12}(\gamma_1, \gamma_2)S_{21}(\gamma_1, \gamma_2),$$

$$K(\gamma_1, \gamma_2) = \frac{1 - |S_{22}(\gamma_1, \gamma_2)|^2 - |S_{11}(\gamma_1, \gamma_2)|^2 + |\Delta_S(\gamma_1, \gamma_2)|^2}{2|S_{12}(\gamma_1, \gamma_2)S_{21}(\gamma_1, \gamma_2)|}.$$

With $\gamma_1 = \gamma_2 = 0$, $B_L(0, 0)$, $B_S(0, 0)$, $C_L(0, 0)$, $C_S(0, 0)$, $\Delta_S(0, 0)$, and $K(0, 0)$ are identical to our previously defined parameters B_L, B_S, C_L, C_S, Δ_S and K respectively.

By the conversion expressions of the last section, we can verify the following identities between the above parameters in different reference systems. As before $\eta_i = (1 - \gamma_i) / (1 - \gamma_i^*)$ with $i = 1$ or 2.

$$B_L(\gamma_1, 0) = B_L, \tag{6.4.1}$$

$$B_L(\gamma_1, \gamma_2) = B_L(0, \gamma_2) = [B_L + \gamma_2\gamma_2^* B_L - \gamma_2 C_L - \gamma_S^* C_S^*] / (1 - |\gamma_2|^2), \tag{6.4.2}$$

$$B_S(0, \gamma_2) = B_S, \tag{6.4.3}$$

$$B_S(\gamma_1, \gamma_2) = B_S(\gamma_1, 0) = [B_S + \gamma_1\gamma_1^* B_S - \gamma_1 C_S - \gamma_1^* C_S^*] / (1 - |\gamma_1|^2), \tag{6.4.4}$$

$$C_L(\gamma_1, 0) = C_L, \tag{6.4.5}$$

$$C_L(\gamma_1, \gamma_2) = C_L(0, \gamma_2) = \eta_2[C_L + \gamma_2^* \gamma_2^* C_L^* - 2\gamma_2^* B_L] / (1 - |\gamma_2|^2), \tag{6.4.6}$$

$$C_S(0,\gamma_2) = C_S,\tag{6.4.7}$$

$$C_S(\gamma_1,\gamma_2) = C_S(\gamma_1,0) = \eta_1[C_S + \gamma_1^*\gamma_1^*C_S^* - 2\gamma_1^*B_S]/(1-|\gamma_1|^2),\tag{6.4.8}$$

$$B_L(\gamma_1,\gamma_2) + B_S(\gamma_1,\gamma_2) = B_L(0,\gamma_2) + B_S(\gamma_1,0)$$

$$= (1-|\Delta_S(\gamma_1,\gamma_2)|^2)/|S_{12}(\gamma_1,\gamma_2)S_{21}(\gamma_1,\gamma_2)|,\tag{6.4.9}$$

$$|S_{12}(\gamma_1,\gamma_2)|/|S_{21}(\gamma_1,\gamma_2)| = |S_{12}|/|S_{21}|,\tag{6.4.10}$$

$$K(\gamma_1,\gamma_2) = K.\tag{6.4.11}$$

The following relations also hold between B_L, C_L and K and between B_S, C_S and K

$$B_L(\gamma_1,\gamma_2)^2 - C_L(\gamma_1,\gamma_2)C_L(\gamma_1,\gamma_2)^* = K^2 - 1,$$

$$B_S(\gamma_1,\gamma_2)^2 - C_S(\gamma_1,\gamma_2)C_S(\gamma_1,\gamma_2)^* = K^2 - 1.$$

We note that with $K = 1$ and $|\Delta_S| = 1$ and

$$B_L(\gamma_1,\gamma_2) = B_S(\gamma_1,\gamma_2) = B_L = B_S = 0,\tag{6.4.12}$$

$$C_L(\gamma_1,\gamma_2) = C_S(\gamma_1,\gamma_2) = C_L = C_S = 0.\tag{6.4.13}$$

Example (6.4.1)

The scattering parameters of a network in a system of resistive reference impedances $\hat{Z}_1 = \hat{Z}_2 = 50$ are given as

$$S_{11} = 0.80\angle 30°, \qquad S_{12} = 0.05\angle -60°,$$

$$S_{21} = 2.00\angle 0°, \qquad S_{22} = 0.60\angle 45°.$$

Find $B_L(\gamma_1,\gamma_2)$, $B_S(\gamma_1,\gamma_2)$, $|C_L(\gamma_1,\gamma_2)|$, $|C_S(\gamma_1,\gamma_2)|$, $\Delta_S(0,0)$ and K in systems of reference impedances:

(i) $(\hat{Z}_1 = 50, \quad \hat{Z}_2 = 50)$, (ii) $(\hat{Z}_1 = 100 - j50, \quad \hat{Z}_2 = 50)$,

(iii) $(\hat{Z}_1 = 50, \quad \hat{Z}_2 = 75 + j25)$, (iv) $(\hat{Z}_1 = 100 - j50, \quad \hat{Z}_2 = 75 + j25)$.

Show that K is independent of the reference impedance.

Solution

For the given reference impedances we have the reference reflection coefficients as $(\gamma_1 = 0, \quad \gamma_2 = 0)$, $(\gamma_1 = 0.447\angle -26.6°, \gamma_2 = 0)$, $(\gamma_1 = 0, \quad \gamma_2 = 0.277\angle 33.7°)$ and $(\gamma_1 = 0,447\angle -26.6°, \quad \gamma_2 = 0.277\angle 33.7°)$, respectively.

The results of calculations of the required scattering parameters for the given reference reflection coefficients were given in Example (6.3.3). Other parameters can be found by direct substitution and are given below.

(γ_1, γ_2)	$(0, 0)$	$(\gamma_1, 0)$	$(0, \gamma_2)$	(γ_1, γ_2)
$B_L(\gamma_1, \gamma_2)$	2.06	2.06	1.88	1.88
$B_S(\gamma_1, \gamma_2)$	4.86	2.02	4.86	2.02
$\|C_L(\gamma_1, \gamma_2)\|$	1.69	1.69	1.47	1.47
$\|C_S(\gamma_1, \gamma_2)\|$	4.72	1.64	4.72	1.64
$\Delta_S(\gamma_1, \gamma_2)$	$0.555\angle 82.3°$	$0.460\angle 130.8°$	$0.595\angle 91.6°$	$0.505\angle 135.3°$
K	1.54	1.54	1.54	1.54

As it can be seen, K is independent of the reference impedances and the following relations also always hold:

$$B_L(\gamma_1, \gamma_2)^2 - C_L(\gamma_1, \gamma_2)C_L(\gamma_1, \gamma_2)^* = K^2 - 1,$$

$$B_S(\gamma_1, \gamma_2)^2 - C_S(\gamma_1, \gamma_2)C_S(\gamma_1, \gamma_2)^* = K^2 - 1,$$

$$B_L(\gamma_1, \gamma_2) + B_S(\gamma_1, \gamma_2) = (1 - |\Delta_S(\gamma_1, \gamma_2)|^2)/|S_{12}(\gamma_1, \gamma_2)S_{21}(\gamma_1, \gamma_2)|.$$

6.5 The Wave Amplitudes of a General 2-port Network

Consider a 2-port network, connecting a generator of emf E_S and internal impedance Z_S to a load of impedance Z_L, as shown in Fig. (6.5.1). With the reference impedances $\hat{Z}_1$ and $\hat{Z}_2$ taken as Z_S and Z_L respectively, the reference reflection coefficients can be denoted by Γ_S and Γ_L and are given as

$$\Gamma_S = \gamma_1 = \frac{Z_S - R_{01}}{Z_S + R_{01}}, \qquad \Gamma_L = \gamma_2 = \frac{Z_L - R_{02}}{Z_L + R_{02}}.$$

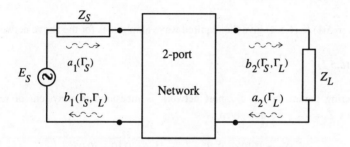

Fig. (6.5.1)

From the definition of the scattering parameters, we can write the relation between $b_1(\Gamma_S, \Gamma_L), b_2(\Gamma_S, \Gamma_L)$ and $a_1(\Gamma_S)$ and $a_2(\Gamma_L)$ as

$$b_1(\Gamma_S, \Gamma_L) = S_{11}(\Gamma_S, \Gamma_L)a_1(\Gamma_S) + S_{12}(\Gamma_S, \Gamma_L)a_2(\Gamma_L),$$

$$b_2(\Gamma_S, \Gamma_L) = S_{21}(\Gamma_S, \Gamma_L)a_1(\Gamma_S) + S_{22}(\Gamma_S, \Gamma_L)a_2(\Gamma_L)$$

where

$$a_1(\Gamma_S) = \frac{E_S}{2\sqrt{R_S}}$$

and with no generator connected to port 2, $a_2(\Gamma_L) = 0$.

Hence defining e_S

$$e_S = \frac{E_S}{2\sqrt{R_S}},$$

the above equation can be written as

$$a_1(\Gamma_S) = e_S, \tag{6.5.1}$$

$$a_2(\Gamma_L) = 0, \tag{6.5.2}$$

$$b_1(\Gamma_S, \Gamma_L) = e_S S_{11}(\Gamma_S, \Gamma_L), \tag{6.5.3}$$

$$b_2(\Gamma_S, \Gamma_L) = e_S S_{21}(\Gamma_S, \Gamma_L). \tag{6.5.4}$$

Equations (6.5.1) to (6.5.4) give the required wave amplitudes for the above network.

Example (6.5.1)

The scattering parameters of a 2-port network is measured in a system of reference impedances $\widehat{Z}_1 = \widehat{Z}_2 = 50$ and given as

$$S_{11} = 0.40 + j0.20, \qquad S_{12} = 0.10 + j0.05,$$

$$S_{21} = 3.00 - j0.50, \qquad S_{22} = 0.30 - j0.10.$$

The network is terminated at its input port by a generator of emf 5 and internal impedance $30 + j20$ and at its output port by a load of 70-j30. Calculate $a_1(\Gamma_S), a_2(\Gamma_L)$, $b_1(\Gamma_S, \Gamma_L)$ and $b_2(\Gamma_S, \Gamma_L)$. Find a_1, a_2, b_1 and b_2 and verify that

$$a_1 a_1^* - b_1 b_1^* = a_1(\Gamma_S, \Gamma_L) a_1^*(\Gamma_S, \Gamma_L) - b_1(\Gamma_S, \Gamma_L) b_1^*(\Gamma_S, \Gamma_L),$$

$$a_2 a_2^* - b_2 b_2^* = a_2(\Gamma_S, \Gamma_L) a_2^*(\Gamma_S, \Gamma_L) - b_2(\Gamma_S, \Gamma_L) b_2^*(\Gamma_S, \Gamma_L).$$

Solution

$\Gamma_S = -0.176 + j0.294$ and $\Gamma_L = 0.216 - j0.196$ and hence

$$S_{11}(\Gamma_S, \Gamma_L) = 0.69 + j0.14 \text{ and } S_{21}(\Gamma_S, \Gamma_L) = 2.48 - j0.36.$$

With

$$e_S = \frac{E_S}{2\sqrt{R_S}} = 0.456$$

and using (6.5.1) to (6.5.4) we have

$$a_1(\Gamma_S) = 0.46, \qquad b_1(\Gamma_S, \Gamma_L) = 0.32 + j0.06,$$

$$a_2(\Gamma_S) = 0, \qquad b_2(\Gamma_S, \Gamma_L) = 1.13 + j0.17.$$

The wave amplitudes for the case of resistive references were given in Example (2.4.1) and are

$$a_1 = 0.37 - j0.05, \qquad b_1 = 0.19 + j0.03,$$

$$a_2 = 0.15 - j0.32, \qquad b_2 = 1.11 - j0.46.$$

By substituting we can verify that

$$a_1 a_1^* - b_1 b_1^* = a_1(\Gamma_S)a_1(\Gamma_L)^* - b_1(\Gamma_S, \Gamma_L)b_1(\Gamma_S, \Gamma_L)^* = 0.105,$$

$$a_2 a_2^* - b_2 b_2^* = a_2(\Gamma_S)a_2(\Gamma_L)^* - b_2(\Gamma_S, \Gamma_L)b_2(\Gamma_S, \Gamma_L)^* = -1.31$$

as required.

6.6 Input and Output Reflection Coefficients - Impedance Matching

In terms of generalized scattering parameters, the input and output reflection coefficients can take a very simple form. This can be realized by taking the reference reflection coefficients as ($\gamma_1 = 0$ and $\gamma_2 = \Gamma_L$) and ($\gamma_1 = \Gamma_S$ and $\gamma_2 = 0$) respectively.

With these particular reference impedances the input and output reflection coefficients can be written as

$$\Gamma_{IN} = S_{11}(0, \Gamma_L), \tag{6.6.1}$$

$$\Gamma_{OT} = S_{22}(\Gamma_S, 0) \tag{6.6.2}$$

which can be seen by comparing (6.3.8a) and (6.3.7d) with (3.3.1a) and (3.3.2a) respectively.

From (6.3.6a) and (6.3.5d) we can write

$$S_{11}(\Gamma_S, \Gamma_L) = \eta_S \frac{S_{11}(0, \Gamma_L) - \Gamma_S^*}{1 - \Gamma_S S_{11}(0, \Gamma_L)},$$

$$S_{22}(\Gamma_S, \Gamma_L) = \eta_L \frac{S_{22}(0, \Gamma_S) - \Gamma_L^*}{1 - \Gamma_L S_{22}(\Gamma_S, 0)}.$$

Hence in terms of $S_{11}(\Gamma_S, \Gamma_L)$ and $S_{22}(\Gamma_S, \Gamma_L)$ the input and output reflection coefficients can be written as

$$\Gamma_{IN} = \frac{S_{11}(\Gamma_S, \Gamma_L) + \eta_S \Gamma_S^*}{\eta_S + \Gamma_S S_{11}(\Gamma_S, \Gamma_L)}, \tag{6.6.3}$$

$$\Gamma_{OT} = \frac{S_{22}(\Gamma_S, \Gamma_L) + \eta_L \Gamma_L^*}{\eta_L + \Gamma_L S_{22}(\Gamma_S, \Gamma_L)} \tag{6.6.4}$$

respectively. η_S and η_L are similarly defined as η_1 and η_2.

Input port and output port and simultaneous conjugate matching

Conjugate matching is best discussed in terms of the scattering parameters given in a system of resistive references, as in Chapter 3. However, the following relations expressed in terms of the generalized scattering parameters are of interest.

For the source-input and load-output conjugate matching

$$S_{11}(\Gamma_{SM}, \Gamma_L) = 0, \tag{6.6.5}$$

$$S_{22}(\Gamma_S, \Gamma_{LM}) = 0 \tag{6.6.6}$$

respectively. (6.6.5) and (6.6.6) can be derived by setting $\Gamma_S = \Gamma_{SM}$, $\Gamma_{IN} = \Gamma_{SM}^*$ in (6.6.3) and $\Gamma_L = \Gamma_{LM}$ and $\Gamma_{OT} = \Gamma_{LM}^*$ in (6.6.4).

For the input and output simultaneous conjugate matching (6.6.5) and (6.6.6) are simultaneously satisfied and hence

$$S_{11}(\Gamma_{SM}, \Gamma_{LM}) = 0, \tag{6.6.7a}$$

$$S_{22}(\Gamma_{SM}, \Gamma_{LM}) = 0. \tag{6.6.7b}$$

It may be of interest to note that in this reference system K has the simple form

$$K = \frac{1 + |S_{12}(\Gamma_{SM}, \Gamma_{LM})S_{21}(\Gamma_{SM}, \Gamma_{LM})|^2}{2|S_{12}(\Gamma_{SM}, \Gamma_{LM})S_{21}(\Gamma_{SM}, \Gamma_{LM})|}. \tag{6.6.8}$$

Example (6.6.1)

The scattering parameters of the network of Examples (3.4.1) to (3.4.3) were given as

$$S_{11} = 0.6\angle 30° \qquad S_{12} = 0.02\angle 15°$$

$$S_{21} = 2.0\angle 0° \qquad S_{22} = 0.8\angle -45°$$

With a load of $Z_L = 50 + j25$, the required source reflection coefficient for the input conjugate matching was found to be $\Gamma_{SM} = \Gamma_{IN}{}^* = 0.518 - j0.311$, verify that $S_{11}(\Gamma_{SM}, \Gamma_L) = 0$. Similarly with a source of impedance $Z_S = 30 - j45$, the load reflection coefficient for the output port conjugate matching was given as $\Gamma_{LM} = \Gamma_{OT}{}^* = 0.567 + j0.590$, verify that $S_{22}(\Gamma_S, \Gamma_{LM}) = 0$. Finally for the simultaneous conjugate matching the required source and load reflection coefficients were found as
$\Gamma_{SM} = 0.578 - j0.386$ and $\Gamma_{LM} = 0.610 + j0.583$, verify that

$$S_{11}(\Gamma_{SM}, \Gamma_{LM}) = 0,$$

$$S_{22}(\Gamma_{SM}, \Gamma_{LM}) = 0.$$

Solution

The above relations can be verified by direct substitution of the relevant parameters to (6.3.3a) and (6.3.3d).

6.7 Stability Considerations

For a particular source and load impedance, a network is stable at its input if

$$Re(Z_S + Z_{IN}) > 0 \tag{6.7.1}$$

and at its output if

$$\text{Re}(Z_L + Z_{OT}) > 0. \tag{6.7.2}$$

As we have discussed before, for a bilateral network, when the input port oscillates the output port also oscillates. For stable network therefore, (6.7.1) and (6.7.2) are both satisfied.

From the expression

$$S_{11}(\Gamma_S, \Gamma_L) = \frac{Z_{IN} - Z_S^*}{Z_{IN} + Z_S},$$

we can write

$$Z_{IN} = \frac{Z_S S_{11}(\Gamma_S, \Gamma_L) + Z_S^*}{1 - S_{11}(\Gamma_S, \Gamma_L)}$$

or

$$Z_S + Z_{IN} = \frac{Z_S + Z_S^*}{1 - S_{11}(\Gamma_S, \Gamma_L)}.$$

Hence for the real part of $Z_S + Z_{IN}$ to be greater than zero, with the source impedance passive, we require

$$\text{Re}[S_{11}^*(\Gamma_S, \Gamma_L)] = \text{Re}[S_{11}(\Gamma_S, \Gamma_L)] < 1. \tag{6.7.3}$$

Similarly for stability for particular values of Z_S and Z_L, the following relation holds

$$\text{Re}[S_{22}(\Gamma_S, \Gamma_L)] < 1. \tag{6.7.4}$$

Stability independent of the source or the load impedance

As we have discussed in Chapter 4, we can consider the network stability for all source impedances. Similarly we can consider the stability of a network for all load impedances. The stability circles define the boundaries of the stable and potentially unstable regions in the Γ_L-plane and Γ_S-plane respectively. The center and radius of the Γ_L-plane stability circle were given by (4.2.5) and (4.2.6), or alternatively by (4.2.9) and (4.2.10) in terms of B_L and K. From (6.4.1) and (6.4.9) we have $B_L = B_L(\Gamma_S, 0)$ and $K = K(\Gamma_S, 0)$. Hence the expressions for the center and the radius of the stability circles in Γ_L-plane are valid if we substitute $\mathbf{S}(\Gamma_S, 0)$-parameters for the $\mathbf{S}$-parameters in the above expressions. Similarly the expressions for the center and radius of the stability circle in Γ_S-plane as given by (4.2.7) and (4.2.8), are still valid if $\mathbf{S}(0, \Gamma_L)$-parameters are substituted for the $\mathbf{S}$-parameters.

Unconditional stability

For unconditional stability we had the requirements of

$$K > 1 \quad \text{and} \quad |\Delta_S| < 1.$$

In Chapter 3 we have shown that when $K > 1$ and $|\Delta_S| < 1$, then $B_L > |C_L|$ and $B_S > |C_S|$. From (6.4.2) and (6.4.4), it is clear that when these inequalities are satisfied, the inequalities $B_L(0, \gamma_2) > 0$ and $B_S(\gamma_1, 0) > 0$ are also satisfied. The reverse of this statement is also true.

Hence with

$$(1 - |\Delta_S(\gamma_1, \gamma_2)|^2) / |S_{12}(\gamma_1, \gamma_2) S_{21}(\gamma_1, \gamma_2)| = B_L(0, \gamma_2) + B_S(\gamma_1, 0)$$

as given by (6.4.10), we conclude that a network is unconditionally stable if the relations

$$K(\gamma_1, \gamma_2) = K > 1 \quad \text{and} \quad |\Delta_S(\gamma_1, \gamma_2)| < 1$$

are satisfied for any pair of reference reflection coefficients γ_1 and γ_2.

6.8 Definitions and Expressions for Power in a 2-port Network

Incident, reflected and delivered powers

Consider a 2-port network connected at port 1 to a source of emf E_1 and impedance Z_{S1}, and at port 2 to a source of emf E_2 and impedance Z_{S2}, as shown in Fig. (6.8.1). Current through and voltage across port 1 is denoted by I_1 and V_1 and similarly by I_2 and V_2 for port 2.

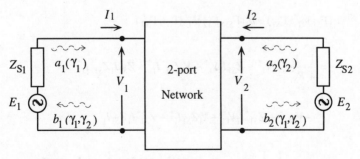

Fig. (6.8.1)

Taking the reference impedances as $\hat{Z}_1$ and $\hat{Z}_2$ we define

$$a_1(\gamma_1,\gamma_2)a_1^*(\gamma_1,\gamma_2)$$

as the *incident power* on port 1 and

$$a_2(\gamma_1,\gamma_2)a_2^*(\gamma_1,\gamma_2)$$

as the *incident power* on port 2.

Similarly we define

$$b_1(\gamma_1,\gamma_2)b_1^*(\gamma_1,\gamma_2)$$

as the *reflected power* from port 1 and

$$b_2(\gamma_1,\gamma_2)b_2^*(\gamma_1,\gamma_2)$$

as the *reflected power* from port 2.

The difference between the incident power on port 1 and the reflected power from the same port

$$a_1(\gamma_1,\gamma_2)a_1^*(\gamma_1,\gamma_2)-b_1(\gamma_1,\gamma_2)b_1^*(\gamma_1,\gamma_2) \tag{6.8.1}$$

is the power delivered to port 1 of the network. Similarly the difference

$$a_2(\gamma_1,\gamma_2)a_2^*(\gamma_1,\gamma_2)-b_2(\gamma_1,\gamma_2)b_2^*(\gamma_1,\gamma_2) \tag{6.8.2}$$

is the power delivered to port 2 of the network.

Expressions (6.8.1) and (6.8.2) correctly express the power delivered to each port, irrespective of the chosen reference impedances. For the case of $\hat{Z}_1 = Z_{S1}$ ($\gamma_1 = \Gamma_{S1}$) and $\hat{Z}_2 = Z_{S2}$ ($\gamma_2 = \Gamma_{S2}$) the fact that (6.8.1) and (6.8.2) represent the delivered power to port 1 and port 2 of the network respectively, can be verified easily using (6.1.15) and (6.1.16). Making the appropriate substitutions we can write

$$a_1(\Gamma_{S1})a_1^*(\Gamma_{S1})-b_1(\Gamma_{S1},\Gamma_{S2})b_1^*(\Gamma_{S1},\Gamma_{S2})$$

$$=\frac{1}{4}\frac{1}{R_{S1}}(V_1V_1^*+Z_{S1}I_1V_1^*+V_1Z_{S1}^*I_1^*+Z_{S1}I_1Z_{S1}^*I_1^*$$

$$-V_1V_1^*+Z_{S1}^*I_1V_1^*+V_1Z_{S1}I_1^*-Z_{S1}^*I_1Z_{S1}I_1^*)$$

$$=\frac{1}{4}\frac{1}{R_{S1}}(Z_{S1}+Z_{S1}^*)(I_1V_1^*+V_1I_1^*)=\frac{1}{2}(I_1V_1^*+V_1I_1^*). \tag{6.8.3a}$$

Similarly,

$$a_2(\Gamma_2)a_2^{\ *}(\Gamma_2) - b_2(\Gamma_1,\Gamma_2)b_2^{\ *}(\Gamma_1,\Gamma_2)$$

$$= \frac{1}{2}(I_2 V_2^{\ *} + V_2 I_2^{\ *}). \tag{6.8.3b}$$

The right-hand sides of (6.8.3a) and (6.8.3b) are clearly the power delivered to ports 1 and 2 of the network respectively.

Power definitions for a network connected to a single generator

We now apply the above general power expressions to the situation where we have a single generator of emf E_S and impedance Z_S connected to port 1 or the input port of a 2-port network. Port 2 or the output port is terminated by an impedance of Z_L. The circuit is shown in Fig. (6.8.2).

For this particular situation we can define the following power expressions for the reference impedance $\hat{Z}_1$ equal to the source impedance Z_S, and the reference impedance $\hat{Z}_2$ equal to the load impedance Z_L. For the case of conjugate matching, source and load impedances are the impedances required for the conjugate matching condition.

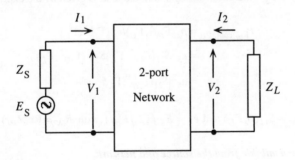

Fig. (6.8.2)

(1) Input power

Input power P_{IN} is the power delivered from the source to the network at its input port. From (6.8.1)

$$P_{IN} = a_1(\Gamma_S)a_1^{\ *}(\Gamma_S) - b_1(\Gamma_S,\Gamma_L)b_1^{\ *}(\Gamma_S,\Gamma_L)$$

$$= e_S^{\ 2}[1 - |S_{11}(\Gamma_S,\Gamma_L)|^2]. \tag{6.8.4}$$

(2) Power available from the source

Power available from the source P_{AVS} is the power delivered from the source to the network at its input port, when the input impedance of the network is a conjugate match of the source impedance.

With $b_1(\Gamma_S, \Gamma_{IN} = \Gamma_S{}^*) = 0$, from (6.8.4)

$$P_{AvS} = a_1(\Gamma_S) a_1{}^*(\Gamma_S) = e_S{}^2. \tag{6.8.5}$$

(3) Power dissipated in the load

Power dissipated in the load P_L can be written as

$$\begin{aligned}
P_L &= b_2(\Gamma_S, \Gamma_L) b_2{}^*(\Gamma_S, \Gamma_L) \\
&= e_S{}^2 |S_{21}(\Gamma_S, \Gamma_L)|^2. \tag{6.8.6}
\end{aligned}$$

(4) Power dissipated in the load with input impedance a conjugate match of the source impedance.

With $\Gamma_{IN} = \Gamma_S{}^*$, the power dissipated in the load is given by the expression

$$\begin{aligned}
P_{\Gamma_{IN} \text{ conj. } \Gamma_S} &= b_2(\Gamma_{IN}{}^*, \Gamma_L) b_2{}^*(\Gamma_{IN}{}^*, \Gamma_L) \\
&= e_S{}^2 S_{21}(\Gamma_{IN}{}^*, \Gamma_L) S_{21}{}^*(\Gamma_{IN}{}^*, \Gamma_L) \\
&= e_S{}^2 |S_{21}(\Gamma_{IN}{}^*, \Gamma_L)|^2 \tag{6.8.7}
\end{aligned}$$

where $\Gamma_{IN} = S_{11}(0, \Gamma_L)$ and $e_S = E_S / (2\sqrt{R_{IN}})$ with $R_{IN} = \text{Re}(Z_{IN})$.

(5) Power available from the source and network

Power available from the source and network P_{AvSN} is the power delivered to the load when the load impedance is a conjugate match to the combined impedance of the source and network.

With $\Gamma_L = \Gamma_{OT}{}^*$

$$\begin{aligned}
P_{AvSN} &= b_2(\Gamma_S, \Gamma_{OT}{}^*) b_2{}^*(\Gamma_S, \Gamma_{OT}{}^*) \\
&= e_S{}^2 |S_{21}(\Gamma_S, \Gamma_{OT}{}^*)|^2 \tag{6.8.8}
\end{aligned}$$

where $\Gamma_{OT} = S_{22}(\Gamma_S, 0)$.

(6) Maximum available power

Maximum available power P_{MAX} is the power delivered to the load from the source and network in the case of simultaneous conjugate matching. Hence

$$P_{MAX} = b_2(\Gamma_{IN}^*, \Gamma_{OT}^*) b_2^*(\Gamma_{IN}^*, \Gamma_{OT}^*)$$

$$= e_S^2 S_{21}(\Gamma_{IN}^*, \Gamma_{OT}^*) S_{21}^*(\Gamma_{IN}^*, \Gamma_{OT}^*)$$

$$= e_S^2 \left| S_{21}(\Gamma_{SMAX}, \Gamma_{LMAX}) \right|^2 \qquad (6.8.9)$$

where Γ_{SMAX} and Γ_{LMAX} are the source and load reflection coefficients for simultaneous matching, as defined in section (3.1). In this case $e_S = E_S / (2\sqrt{R_{SMAX}})$ with $R_{SMAX} = \text{Re}(Z_{SMAX})$.

(7)Power delivered to the network

Power delivered to the network P_{NWK}, is the difference beween the input power to the network and the power delivered to the load. Hence

$$P_{NWK} = P_{IN} - P_L$$

$$= e_S^2 [1 - S_{11}(\Gamma_S, \Gamma_L) S_{11}^*(\Gamma_S, \Gamma_L) - S_{21}(\Gamma_S, \Gamma_L) S_{21}^*(\Gamma_S, \Gamma_L)]$$

$$= e_S^2 [1 - \left| S_{11}(\Gamma_S, \Gamma_L) \right|^2 - \left| S_{21}(\Gamma_S, \Gamma_L) \right|^2]. \qquad (6.8.10)$$

The above expressions should be evaluated by first calculating the required scattering parameters in systems of appropriate references. The power terms then can readily be found by substitution.

Power gains in terms of the scattering parameters with references equal to the source and load impedances.

From the expressions for the different power terms given in the previous section, the various power gains defined in Chapter 5 can readily be written as:

(i)Transducer power gain

$$G_T = \frac{P_L}{P_{AvS}} = \left| S_{21}(\Gamma_S, \Gamma_L) \right|^2, \qquad (6.8.11)$$

(ii) Operating power gain

$$G_P = \frac{P_L}{P_{IN}} = \frac{\left|S_{21}(\Gamma_S, \Gamma_L)\right|^2}{1 - \left|S_{11}(\Gamma_S, \Gamma_L)\right|^2},$$ (6.8.12)

(iii) Available power gain

$$G_{AV} = \frac{P_{AvSN}}{P_{AvS}} = \left|S_{21}(\Gamma_S, \Gamma_{OT}{}^*)\right|^2,$$ (6.8.13)

(iv) Maximum available power gain

$$G_{MAX} = \frac{P_{MAX}}{P_{AvS}} = \left|S_{21}(\Gamma_{SMAX}, \Gamma_{LMAX})\right|^2.$$ (6.8.14)

Alternatively from (5.2.4b) and (6.4.11),

$$G_{MAX} = \frac{\left|S_{21}(\Gamma_{SMAX}, \Gamma_{LMAX})\right|}{\left|S_{12}(\Gamma_{SMAX}, \Gamma_{LMAX})\right|}[K - \sqrt{(K^2 - 1)}]$$

which can be verified using (6.8.14) and (6.6.8).

Example (6.8.1)

Calculate all the power terms in Example (5.1.1), using the generalized scattering parameters with the reference impedances equal to the impedances of the source and the load. Assume that the total available power from the source remains unchanged in all cases.

Solution

(1) Input power

$$P_{IN} = e_S{}^2[1 - \left|S_{11}(\Gamma_S, \Gamma_L)\right|^2]$$

From the scattering parameters conversion $S_{11}(\Gamma_S, \Gamma_L) = 0.69 + j0.14$ and $e_S{}^2 = \dfrac{E_S{}^2}{4R_S} = 0.208$ and hence $P_{IN} = 0.208(1 - 0.496) = 0.105$.

(2) Power available from the source

$$P_{AVS} = e_S{}^2 = \frac{E_S{}^2}{4R_S} = \frac{25}{120} = 0.208$$

(3) Power dissipated in the load

$$P_L = e_S{}^2 \left| S_{21}(\Gamma_S, \Gamma_L) \right|^2$$

In this case $S_{21}(\Gamma_S, \Gamma_L) = 2.48 - j0.36$ and hence $|S_{21}(\Gamma_S, \Gamma_L)|^2 = 6.28$ and $P_L = 0.208 \times 6.28 = 1.31$.

(4) Power dissipated in the load with input impedance conjugate match of the source impedance

$$P_{\Gamma_{IN} \text{ conj. } \Gamma_S} = e_S{}^2 \left| S_{21}(\Gamma_{IN}{}^*, \Gamma_L) \right|^2$$

and with $S_{21}(\Gamma_{IN}{}^*, \Gamma_L) = 3.40 + j0.97$, $|S_{21}(\Gamma_{IN}{}^*, \Gamma_L)|^2 = 12.5$, $P_L = 0.208 \times 12.5 = 2.60$

(5) Power available from the source and network

$$P_{AVN} = e_S{}^2 \left| S_{21}(\Gamma_S, \Gamma_{OUT}{}^*) \right|^2$$

With $S_{21}(\Gamma_S, \Gamma_{OUT}{}^*) = 2.40 + j0.97$ or $|S_{21}(\Gamma_S, \Gamma_{OUT}{}^*)| = 6.7$, $P_{AVN} = 0.208 \times 6.7 = 1.40$.

(6) Maximum available power

$$P_{MAX} = e_S{}^2 \left| S_{21}(\Gamma_{SMAX}, \Gamma_{LMAX})^2 \right|$$

With $S_{21}(\Gamma_{SMAX}, \Gamma_{LMAX}) = 4.20 + j0.50$ and hence $|S_{21}(\Gamma_{SMAX}, \Gamma_{LMAX})|^2 = 17.9$, $P_{MAX} = 0.208 \times 17.9 = 3.72$.

(7) Power delivered to the network

$$P_{NWK} = P_{IN} - P_L = 0.105 - 1.31 = -1.21$$

Hence power is generated (or added) inside the network.

6.9 Power Consideration of Special 2-port Networks

With relations between different parameters investigated in Section 4 of this Chapter (Equations (6.4.1) to (6.4.11)), it can be verified that the conditions for a network to be passive, lossless or reciprocal have identical forms as derived in Section 5.5. Hence to find these conditions in terms of the generalized scattering parameters, it is sufficient to formally replace the scattering parameters S_{11}, S_{12}, S_{21} and S_{22} by scattering parameters $S_{11}(\gamma_1,\gamma_2)$, $S_{21}(\gamma_1,\gamma_2)$, $S_{22}(\gamma_1,\gamma_2)$ and $S_{12}(\gamma_1,\gamma_2)$ respectively. The reference impedances γ_1 and γ_2 can assume any arbitrary value. The result of this substitution is summarized below.

Passive network

For a passive network we require $P_{NWK} \geq 0$ for all source and load impedances. If a network is passive, the following inequalities hold between the generalized scattering parameters.

$$K(\gamma_1,\gamma_2) > 1, \tag{6.9.1}$$

$$|\Delta_S(\gamma_1,\gamma_2)| < 1, \tag{6.9.2}$$

$$|S_{11}(\gamma_1,\gamma_1)|^2 + |S_{21}(\gamma_1,\gamma_1)|^2$$

$$+ |S_{22}(\gamma_1,\gamma_1)|^2 + |S_{12}(\gamma_1,\gamma_1)|^2 < 1 + |\Delta_S(\gamma_1,\gamma_1)|^2 \tag{6.9.3}$$

and

$$1 - |S_{11}(\gamma_1,\gamma_2)|^2 - |S_{21}(\gamma_1,\gamma_2)|^2 > 0, \tag{6.9.4a}$$

$$1 - |S_{22}(\gamma_1,\gamma_2)|^2 - |S_{12}(\gamma_1,\gamma_2)|^2 > 0. \tag{6.9.4b}$$

Lossless 2-port network

For a passive network we require $P_{NWK} = 0$, for all source and load impedances. If a network is lossless, the following identities are satisfied:

$$|\Delta_S(\gamma_1,\gamma_2)| = 1, \tag{6.9.5}$$

$$|S_{22}(\gamma_1,\gamma_2)| = |S_{11}(\gamma_1,\gamma_2)|, \tag{6.9.6a}$$

$$|S_{21}(\gamma_1,\gamma_2)| = |S_{12}(\gamma_1,\gamma_2)|, \tag{6.9.6b}$$

$$B_L(\gamma_1,\gamma_2) = B_S(\gamma_1,\gamma_2) = 0, \tag{6.9.7}$$

$$|C_L(\gamma_1,\gamma_2)| = |C_S(\gamma_1,\gamma_2)| = 0, \tag{7.9.8}$$

$$1 - |S_{11}(\gamma_1,\gamma_2)|^2 - |S_{21}(\gamma_1,\gamma_2)|^2 = 0, \tag{6.9.9a}$$

$$\tag{6.9.9b}$$
$$1 - |S_{22}(\gamma_1,\gamma_2)|^2 - |S_{12}(\gamma_1,\gamma_2)|^2 = 0.$$

In addition, the phases of the generalized scattering parameters satisfy the relation

$$\phi_{11}(\gamma_1,\gamma_2) + \phi_{22}(\gamma_1,\gamma_2) - \phi_{12}(\gamma_1,\gamma_2) - \phi_{21}(\gamma_1,\gamma_2) = \pi$$

where $\phi_{11}(\gamma_1,\gamma_2), \phi_{12}(\gamma_1,\gamma_2), \phi_{21}(\gamma_1,\gamma_2)$ and $\phi_{22}(\gamma_1,\gamma_2)$ represent the phases of $S_{11}(\gamma_1,\gamma_2)$, $S_{12}(\gamma_1,\gamma_2)$, $S_{21}(\gamma_1,\gamma_2)$ and $S_{22}(\gamma_1,\gamma_2)$ respectively.

Reciprocal 2-port networks

For a reciprocal network we have $S_{12} = S_{21}$, which from (6.3.3b) and (6.3.3c) leads to the condition

$$S_{12}(\gamma_1,\gamma_2) = S_{21}(\gamma_1,\gamma_2).$$

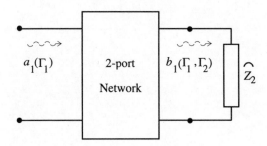

Fig. (6.9.1a) $a_2(\Gamma_2) = 0$

For a reciprocal network if (a) a generator is connected to port 1 of the network and port 2 is terminated by its reference impedance $\hat{Z}_2$ (Fig. (6.9.1a)) and (b) a generator is connected to port 2 of the network and port 2 is terminated by its reference impedance $\hat{Z}_1$ (Fig. (6.9.1b)), then with $a_1(\Gamma_1) = a_2(\Gamma_2)$, we have $b_1(\Gamma_1,\Gamma_2) = b_2(\Gamma_1,\Gamma_2)$.

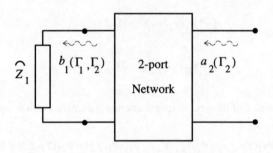

Fig. (6.9.1b) $a_1(\Gamma_1) = 0$

Chapter 7

The Scattering Parameters of an *N*-port Network

In this Chapter, we generalize the definition of the scattering parameters to networks with an arbitrary number of ports. As for the case of 2-port networks discussed in the last Chapter, we define the scattering parameters in terms of a set of generalized reference impedances. The expressions for the conversion of the measurable parameters to the generalized scattering parameters are given. The powers associated with the network and the network stability is examined. For any two specific ports of a network, an *N*-port network can be reduced to an equivalent 2-port network. The question of this reduction is discussed and the scattering parameters of the equivalent 2-port network is derived. The possible reverse process is also considered. Finally we discuss three special classes of networks, namely, passive, lossless and reciprocal.

7.1 The Scattering Matrix of an *N*-port Network

Consider an *N*-port network connected to N generators of emf $E_1, .. E_j, ..$ and E_N in series with impedances $\hat{Z}_1, .. \hat{Z}_j, ..$ and $\hat{Z}_N$, taken to be equal to the designated reference impedances, as shown in Fig. (7.1.1).

Fig. (7.1.1)

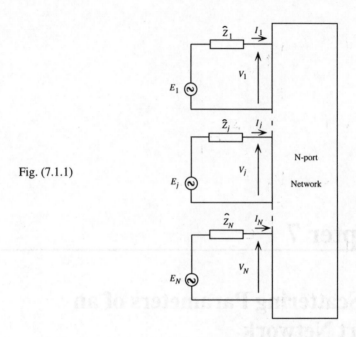

The current flowing through the *j*th port is denoted by I_j and voltage across the same port as V_j. The incident current matrix $\mathbf{I}^i(\gamma_1 .. \gamma_j .. \gamma_N)$ is defined as a column matrix with the *j*th element as

$$I^i{}_j(\gamma_1 .. \gamma_j .. \gamma_N) = \frac{E_j}{\hat{Z}_j + \hat{Z}_j{}^*} = \frac{E_j}{2\hat{R}_j} \tag{7.1.1}$$

where $\hat{R}_j = \mathrm{Re}(\hat{Z}_j)$. This current is the current that would flow with the *j*th generator and the associated impedance disconnected from the corresponding port and terminated by a load equal to the *conjugate* of the reference impedance.

In (7.1.1) the jth port reference reflection coefficient γ_j is given as

$$\gamma_j = \frac{\hat{Z}_j - R_{0j}}{\hat{Z}_j + R_{0j}}$$

where R_{0j} is similarly defined as R_{01} and R_{02} for a 2-port network.

We likewise define the incident voltage matrix $\mathbf{V}^i(\gamma_1..\gamma_j..\gamma_N)$ as a column matrix with elements equal to the voltages across the conjugate impedances. Hence for the jth element

$$V^i{}_j(\gamma_1..\gamma_j..\gamma_N) = \frac{\hat{Z}_j^* E_j}{\hat{Z}_j + \hat{Z}_j^*} = \frac{\hat{Z}_j^* E_j}{2\hat{R}_j}. \tag{7.1.2}$$

From (7.1.1) and (7.1.2) we have

$$\mathbf{V}^i(\gamma_1..\gamma_j..\gamma_N) = \hat{\mathbf{Z}}^* \mathbf{I}^i(\gamma_1..\gamma_j..\gamma_N) \tag{7.1.3}$$

where $\hat{\mathbf{Z}}^*$ is a diagonal matrix with diagonal elements $\hat{Z}_1^*,..\hat{Z}_j^*,..$ and $\hat{Z}_N^*$ and all other elements set to zero.

With $\mathbf{I}$ and $\mathbf{V}$ as the current and voltage matrices with elements I_j and V_j, the reflected current and voltage matrices can be written as

$$\mathbf{I}^r(\gamma_1..\gamma_j..\gamma_N) = \mathbf{I}^i(\gamma_1..\gamma_j..\gamma_N) - \mathbf{I}, \tag{7.1.4}$$

$$\mathbf{V}^r(\gamma_1..\gamma_j..\gamma_N) = \mathbf{V} - \mathbf{V}^i(\gamma_1..\gamma_j..\gamma_N). \tag{7.1.5}$$

For the jth port

$$V_j = E_j - \hat{Z}_j I_j.$$

Substituting for I_j and V_j in terms of the jth port incident and reflected currents and voltages, as given respectively by (7.1.4) and (7.1.5), the above expression can be written as

$$V^i{}_j(\gamma_1..\gamma_j..\gamma_N) + V^r{}_j(\gamma_1..\gamma_j..\gamma_N) = E_j - \hat{Z}_j[I^i{}_j(\gamma_1..\gamma_j..\gamma_N) - I^r{}_j(\gamma_1..\gamma_j..\gamma_N)].$$

But from (7.1.3)

$$V^i{}_j(\gamma_1..\gamma_j..\gamma_N) = \hat{Z}_j^* I^i{}_j(\gamma_1..\gamma_j..\gamma_N)$$

and hence

$$V^r{}_j(\gamma_1..\gamma_j..\gamma_N) = E_j - (\hat{Z}_j + \hat{Z}_j{}^*)I^i{}_j(\gamma_1..\gamma_j..\gamma_N) + \hat{Z}_j I^r{}_j(\gamma_1..\gamma_j..\gamma_N).$$

With $E_j - (\hat{Z}_j + \hat{Z}_j{}^*)I^i{}_j(\gamma_1..\gamma_j..\gamma_N) = 0$ as in (7.1.1), we have

$$V^r{}_j(\gamma_1..\gamma_j..\gamma_N) = \hat{Z}_j I^r{}_j(\gamma_1..\gamma_j..\gamma_N). \tag{7.1.6}$$

Written in matrix form we have

$$\mathbf{V}^r(\gamma_1..\gamma_j..\gamma_N) = \hat{\mathbf{Z}}\,\mathbf{I}^r(\gamma_1..\gamma_j..\gamma_N) \tag{7.1.7}$$

where $\hat{\mathbf{Z}}$ is a diagonal matrix with diagonal elements of the reference impedances $\hat{Z}_1,..\hat{Z}_j,..$ and $\hat{Z}_N$.

The relation between $\mathbf{I}^r(\gamma_1..\gamma_j..\gamma_N)$ and $\mathbf{I}^i(\gamma_1..\gamma_j..\gamma_N)$ is a property of the N-port network and can be expressed as

$$\mathbf{I}^r(\gamma_1..\gamma_j..\gamma_N) = \mathbf{S}^I(\gamma_1..\gamma_j..\gamma_N)\mathbf{I}^i(\gamma_1..\gamma_j..\gamma_N). \tag{7.1.8}$$

$\mathbf{S}^I(\gamma_1..\gamma_j..\gamma_N)$ is known as the current scattering matrix of the network with non-diagonal elements $S^I{}_{kj}(\gamma_1..\gamma_j..\gamma_N)$ and diagonal elements $S^I{}_{jj}(\gamma_1..\gamma_j..\gamma_N)$.

Similarly we can write

$$\mathbf{V}^r(\gamma_1..\gamma_j..\gamma_N) = \mathbf{S}^V(\gamma_1..\gamma_j..\gamma_N)\mathbf{V}^i(\gamma_1..\gamma_j..\gamma_N) \tag{7.1.9}$$

with $\mathbf{S}^V(\gamma_1..\gamma_j..\gamma_N)$ known as the voltage scattering matrix.

As the incident and reflected currents are linearly related to the incident and reflected voltages as given by (7.1.3) and (7.1.7), definition of two different scattering matrices is not necessary. As we have seen for the case of 1-port and 2-port networks, a scattering matrix can be defined being linearly related to the above scattering matrices, but resulting in more simplified power expressions for the network. Defining

$$\mathbf{a}(\gamma_1..\gamma_j..\gamma_N) = \left[\sqrt{\hat{R}}\right]\mathbf{I}^i(\gamma_1..\gamma_j..\gamma_N) = \left[\frac{\sqrt{\hat{R}}}{\hat{Z}^*}\right]\mathbf{V}^i(\gamma_1..\gamma_j..\gamma_N) \tag{7.1.10}$$

as the incident wave amplitude matrix, and

$$\mathbf{b}(\gamma_1..\gamma_j..\gamma_N) = \left[\sqrt{R}\right]\mathbf{I}^r(\gamma_1..\gamma_j..\gamma_N) = \left[\frac{\sqrt{R}}{\hat{Z}}\right]\mathbf{V}^r(\gamma_1..\gamma_j..\gamma_N) \qquad (7.1.11)$$

as the reflected wave amplitude matrix, we define the wave amplitude scattering matrix $\mathbf{S}(\gamma_1..\gamma_j..\gamma_N)$ by the expression

$$\mathbf{b}(\gamma_1..\gamma_j..\gamma_N) = \mathbf{S}(\gamma_1..\gamma_j..\gamma_N)\mathbf{a}(\gamma_1..\gamma_j..\gamma_N). \qquad (7.1.12)$$

In (7.1.10) and (7.1.11) $[\sqrt{R}]$ is a diagonal matrix with the jth diagonal elements $\sqrt{R_j}$. Similarly the jth diagonal element of the matrix $[\sqrt{R}/\hat{Z}]$ is $\sqrt{R_j}/\hat{Z}_j$ and the jth diagonal element of the matrix $[\sqrt{R}/\hat{Z}^*]$ is $\sqrt{R_j}/\hat{Z}_j^*$. The non-diagonal elements are all zero.

The elements of the scattering matrix are hence defined as

$$S_{kj}(\gamma_1..\gamma_j..\gamma_N) = \frac{b_k(\gamma_1..\gamma_j..\gamma_N)}{a_j(\gamma_1..\gamma_j..\gamma_N)}\Bigg|_{a_\ell(\gamma_1..\gamma_j..\gamma_N) = 0}$$

$$(\ell = 1,..k,..N, \quad \ell \neq j).$$

Expressed in words, the scattering parameter S_{kj} is the ratio of the reflected wave from the kth port $b_k(\gamma_1..\gamma_j..\gamma_N)$ to that of the incident wave on the jth port $a_j(\gamma_1..\gamma_j..\gamma_N)$, with only the jth source generating and all other ports terminated by their reference impedances. Terminating the lth port $(\ell \neq j)$ by its reference impedance prevents any reflection from the terminating load of that port. Hence if all ports with the possible exception of the jth port are similarly terminated, the jth port reflected wave is reduced to the reflection from the N-port network itself, excluding the additional reflections from all other terminating loads.

Although the incident and reflected waves are dependent on the reference impedances, the expression

$$a_j(\gamma_1..\gamma_j..\gamma_N)a_j^*(\gamma_1..\gamma_j..\gamma_N) - b_j^*(\gamma_1..\gamma_j..\gamma_N)b_j(\gamma_1..\gamma_j..\gamma_N)$$

gives the power delivered to the jth port, independent of the reference system. $a_j(\gamma_1..\gamma_j..\gamma_N)a_j^*(\gamma_1..\gamma_j..\gamma_N)$ can be interpreted as the incident power on the jth port and $b_j(\gamma_1..\gamma_j..\gamma_N)b_j^*(\gamma_1..\gamma_j..\gamma_N)$ as the reflected power from the same port.

$\mathbf{S}(\gamma_1 .. \gamma_j .. \gamma_N)$ in terms of $\mathbf{S}^I(\gamma_1 .. \gamma_j .. \gamma_N)$ and $\mathbf{S}^V(\gamma_1 .. \gamma_j .. \gamma_N)$ is given as

$$\mathbf{S}(\gamma_1 .. \gamma_j .. \gamma_N) = \left[\sqrt{\widehat{R}}\right] \mathbf{S}^I(\gamma_1 .. \gamma_j .. \gamma_N) \left[\frac{1}{\sqrt{\widehat{R}}}\right], \qquad (7.1.13)$$

$$\mathbf{S}(\gamma_1 .. \gamma_j .. \gamma_N) = \left[\frac{\sqrt{\widehat{R}}}{\widehat{Z}}\right] \mathbf{S}^V(\gamma_1 .. \gamma_j .. \gamma_N) \left[\frac{\widehat{Z}^*}{\sqrt{\widehat{R}}}\right]. \qquad (7.1.14)$$

7.2 Calculation of the Scattering Parameters from the Impedance and Admittance Parameters

In this section we define the impedance and admittance parameters and consider the conversion relations between these parameters and the scattering parameters.

The current through and voltage across the different ports of an N-port network are related, respectively, by an impedance or an admittance matrix in the form of

$$\mathbf{V} = \mathbf{Z}\mathbf{I} \qquad (7.2.1)$$

and

$$\mathbf{I} = \mathbf{Y}\mathbf{V} \qquad (7.2.2)$$

with $\mathbf{Y} = \mathbf{Z}^{-1}$.

The matrix element Z_{kj} of $\mathbf{Z}$ is the ratio of the voltage across the kth port to the current through the jth port with all ports except the jth port open-circuited. The matrix element Y_{kj} of $\mathbf{Y}$ is the ratio of the current through the kth port to that of the voltage across the jth port when all other ports except the jth port are short-circuited.

To find the conversion between the scattering parameters and the impedance and admittance parameters, we write $\mathbf{I}$ and $\mathbf{V}$ in terms of the incident and reflected current and voltages as

$$\mathbf{I} = \mathbf{I}^i(\gamma_1 .. \gamma_j .. \gamma_N) - \mathbf{I}^r(\gamma_1 .. \gamma_j .. \gamma_N)$$

and

$$\mathbf{V} = \mathbf{V}^i(\gamma_1 .. \gamma_j .. \gamma_N) + \mathbf{V}^r(\gamma_1 .. \gamma_j .. \gamma_N).$$

Substituting for $\mathbf{I}^i(\gamma_1..\gamma_j..\gamma_N)$, $\mathbf{V}^i(\gamma_1..\gamma_j..\gamma_N)$, $\mathbf{I}^r(\gamma_1..\gamma_j..\gamma_N)$ and $\mathbf{V}^r(\gamma_1..\gamma_j..\gamma_N)$ from (7.1.10) and (7.1.11) into the above equations, we have

$$\mathbf{I} = \left[\frac{1}{\sqrt{\widehat{R}}}\right]\mathbf{a}(\gamma_1..\gamma_j..\gamma_N) - \left[\frac{1}{\sqrt{\widehat{R}}}\right]\mathbf{b}(\gamma_1..\gamma_j..\gamma_N), \tag{7.2.3}$$

$$\mathbf{V} = \left[\frac{\widehat{Z}^*}{\sqrt{\widehat{R}}}\right]\mathbf{a}(\gamma_1..\gamma_j..\gamma_N) + \left[\frac{\widehat{Z}}{\sqrt{\widehat{R}}}\right]\mathbf{b}(\gamma_1..\gamma_j..\gamma_N). \tag{7.2.4}$$

Hence from (7.2.1), (7.2.3) and (7.2.4)

$$\mathbf{a}(\gamma_1..\gamma_j..\gamma_N) = \frac{1}{2}\left[\frac{1}{\sqrt{\widehat{R}}}\right]\{\mathbf{Z} + \widehat{\mathbf{Z}}\}\mathbf{I}$$

$$\mathbf{b}(\gamma_1..\gamma_j..\gamma_N) = \frac{1}{2}\left[\frac{1}{\sqrt{\widehat{R}}}\right]\{\mathbf{Z} - \widehat{\mathbf{Z}}^*\}\mathbf{I}$$

But

$$\mathbf{b}(\gamma_1..\gamma_j..\gamma_N) = \mathbf{S}(\gamma_1..\gamma_j..\gamma_N)\mathbf{a}(\gamma_1..\gamma_j..\gamma_N)$$

or finally

$$\mathbf{S}(\gamma_1..\gamma_j..\gamma_N) = \left[\frac{1}{\sqrt{\widehat{R}}}\right]\{\mathbf{Z} - \widehat{\mathbf{Z}}^*\}\{\mathbf{Z} + \widehat{\mathbf{Z}}\}^{-1}\left[\sqrt{\widehat{R}}\right]. \tag{7.2.5}$$

Using (7.2.5) we can convert the $\mathbf{Z}$ matrix representation of an N-port network to an $\mathbf{S}(\gamma_1..\gamma_j..\gamma_N)$ matrix representation in any system of reference impedances.

Conversely, to find the $\mathbf{Z}$ matrix in terms of $\mathbf{S}(\gamma_1..\gamma_j..\gamma_N)$ matrix, we write (7.2.5) as

$$\left[\frac{1}{\sqrt{\widehat{R}}}\right]\{\mathbf{Z} - \widehat{\mathbf{Z}}^*\} = \mathbf{S}(\gamma_1..\gamma_j..\gamma_N)\left[\frac{1}{\sqrt{\widehat{R}}}\right]\{\mathbf{Z} + \widehat{\mathbf{Z}}\}$$

and after some rearrangement we have finally

$$\mathbf{Z} = \left[\sqrt{\widehat{R}}\right]\{\mathbf{U} - \mathbf{S}(\gamma_1..\gamma_j..\gamma_N)\}^{-1}\left\{\left[\frac{\widehat{Z}^*}{\sqrt{\widehat{R}}}\right] + \mathbf{S}(\gamma_1..\gamma_j..\gamma_N)\left[\frac{\widehat{Z}}{\sqrt{\widehat{R}}}\right]\right\} \tag{7.2.6}$$

which is the required expression. In (7.2.6) $\mathbf{U}$ is a diagonal matrix with all diagonal elements as unity (identity matrix).

Similarly from (7.2.3) and (7.2.4) we can write

$$\mathbf{a}(\gamma_1..\gamma_j..\gamma_N) = \frac{1}{2}\left[\frac{\widehat{Z}}{\sqrt{R}}\right]\{\widehat{\mathbf{Y}} + \mathbf{Y}\}\mathbf{V},$$

$$\mathbf{b}(\gamma_1..\gamma_j..\gamma_N) = \frac{1}{2}\left[\frac{\widehat{Z}^*}{\sqrt{R}}\right]\{\widehat{\mathbf{Y}}^* - \mathbf{Y}\}\mathbf{V}$$

where $\widehat{\mathbf{Y}}$ is a diagonal matrix with diagonal elements

$$\widehat{Y}_j = \frac{1}{\widehat{Z}_j} \quad (j = 1 \text{ to } N).$$

From the above expressions we find

$$\mathbf{S}(\gamma_1..\gamma_j..\gamma_N) = \left[\frac{\widehat{Z}^*}{\sqrt{R}}\right]\{\widehat{\mathbf{Y}}^* - \mathbf{Y}\}\{\widehat{\mathbf{Y}} + \mathbf{Y}\}^{-1}\left[\frac{\sqrt{R}}{\widehat{Z}}\right]. \qquad (7.2.7)$$

To find $\mathbf{Y}$ in terms of $\mathbf{S}$, we write (7.2.7) as

$$\left[\frac{\widehat{Z}^*}{\sqrt{R}}\right]\{\widehat{\mathbf{Y}}^* - \mathbf{Y}\} = \mathbf{S}(\gamma_1..\gamma_j..\gamma_N)\left[\frac{\widehat{Z}}{\sqrt{R}}\right]\{\widehat{\mathbf{Y}} + \mathbf{Y}\}$$

and after some rearrangement we finally have

$$\mathbf{Y} = \left\{\left[\frac{\widehat{Z}^*}{\sqrt{R}}\right] + \mathbf{S}(\gamma_1..\gamma_j..\gamma_N)\left[\frac{\widehat{Z}}{\sqrt{R}}\right]\right\}^{-1}\{\mathbf{U} - \mathbf{S}(\gamma_1..\gamma_j..\gamma_N)\}\left[\frac{1}{\sqrt{R}}\right]. \qquad (7.2.8)$$

From (7.2.6) and (7.2.8), the relation $\mathbf{Y} = \mathbf{Z}^{-1}$ can be easily verified.

Example (7.2.1)

From expressions (7.2.7) and (7.2.8) for $\mathbf{Y}$ to $\mathbf{S}$ and $\mathbf{S}$ to $\mathbf{Y}$ conversion for N-port networks, find the corresponding expressions for a 2-port network as given by (2.3.16) and (2.3.17).

Solution

For a 2-port network, the matrix terms in (7.2.7) are given as

$$\left[\frac{\sqrt{\widehat{R}}}{\widehat{Z}}\right] = \begin{bmatrix} \sqrt{\widehat{R}_1}/\widehat{Z}_1 & 0 \\ 0 & \sqrt{\widehat{R}_2}/\widehat{Z}_2 \end{bmatrix}, \quad \left[\frac{\widehat{Z}^*}{\sqrt{\widehat{R}}}\right] = \begin{bmatrix} \widehat{Z}_1^*/\sqrt{\widehat{R}_1} & 0 \\ 0 & \widehat{Z}_2^*/\sqrt{\widehat{R}_2} \end{bmatrix}$$

and

$$\left\{\widehat{\mathbf{Y}}+\mathbf{Y}\right\}^{-1} = \frac{1}{\Sigma_Y}\begin{bmatrix} Y_{22}+\widehat{Y}_2 & -Y_{12} \\ \\ -Y_{21} & Y_{11}+\widehat{Y}_1 \end{bmatrix}, \quad \widehat{\mathbf{Y}}^*-\mathbf{Y} = \begin{bmatrix} \widehat{Y}_1^*-Y_{11} & -Y_{12} \\ \\ -Y_{21} & \widehat{Y}_2^*-Y_{22} \end{bmatrix}$$

where

$$\Sigma_Y = (Y_{11}+\widehat{Y}_1)\ (Y_{22}+\widehat{Y}_2) - Y_{12}Y_{21}.$$

After matrix multiplication we can find (2.3.14) and subsequently (2.3.16).
For expression (7.2.8) we have

$$\{\mathbf{U}-\mathbf{S}\}\left[\frac{1}{\sqrt{\widehat{R}}}\right] = \begin{bmatrix} \dfrac{1-S_{11}}{\sqrt{\widehat{R}_1}} & -\dfrac{S_{12}}{\sqrt{\widehat{R}_2}} \\ \\ -\dfrac{S_{21}}{\sqrt{\widehat{R}_1}} & \dfrac{1-S_{22}}{\sqrt{\widehat{R}_2}} \end{bmatrix}$$

and

$$\left\{\left[\frac{\widehat{Z}^*}{\sqrt{\widehat{R}}}\right]+\mathbf{S}\left[\frac{\widehat{Z}}{\sqrt{\widehat{R}}}\right]\right\}^{-1} = \begin{bmatrix} \dfrac{\widehat{Z}_1^*+S_{11}Z_1}{\sqrt{\widehat{R}_1}} & \dfrac{\widehat{Z}_2 S_{12}}{\sqrt{\widehat{R}_2}} \\ \\ \dfrac{\widehat{Z}_1 S_{21}}{\sqrt{\widehat{R}_1}} & \dfrac{\widehat{Z}_2^*+S_{22}\widehat{Z}_2}{\sqrt{\widehat{R}_2}} \end{bmatrix}^{-1}$$

$$= \frac{\sqrt{\widehat{R}_1}\sqrt{\widehat{R}_2}\,\widehat{Y}_1\widehat{Y}_2}{\sigma_Y}\begin{bmatrix} \dfrac{(\widehat{Y}_2/\widehat{Y}_2^*)+S_{22}}{\sqrt{\widehat{R}_2}} & -\dfrac{S_{12}}{\sqrt{\widehat{R}_2}\,\widehat{Y}_2} \\ \\ -\dfrac{S_{21}}{\sqrt{\widehat{R}_1}\,\widehat{Y}_1} & \dfrac{(\widehat{Y}_1/\widehat{Y}_1^*)+S_{11}}{\sqrt{\widehat{R}_1}} \end{bmatrix}$$

where

$$\sigma_Y = \left(S_{11} + \frac{Y_1}{Y_1^*} \right) \left(S_{22} + \frac{Y_2}{Y_2^*} \right) - S_{12} S_{21}$$

and leading finally to (2.3.17).

Example (7.2.2)

The impedance matrix for a 3-port network is given as

$$\mathbf{Z} = \begin{bmatrix} 62 + j71 & 32 - j63 & 55 + j36 \\ 52 - j56 & 34 + j25 & 24 + j37 \\ 55 + j83 & 32 + j91 & 34 - j82 \end{bmatrix},$$

find the scattering matrix of the network for a system of reference impedances $\hat{Z}_1 = 64 + j25$, $\hat{Z}_2 = 81 - j50$ and $\hat{Z}_3 = 25 + j75$.

Solution

For finding $\mathbf{S}$ from $\mathbf{Z}$, we use (7.2.5) with the different required matrices given as

$$\left[\sqrt{\hat{R}} \right] = \begin{bmatrix} 8.00 & 0 & 0 \\ 0 & 9.00 & 0 \\ 0 & 0 & 5.00 \end{bmatrix},$$

$$\left\{ \mathbf{Z} + \hat{\mathbf{Z}} \right\}^{-1} = 10^{-3} \times \begin{bmatrix} 3.76 - j3.24 & -0.60 + j4.80 & -2.10 - j1.10 \\ -1.44 + j3.83 & 3.53 - j1.40 & 1.85 - j4.12 \\ -1.13 - j2.25 & 4.17 - j7.84 & 9.48 + j4.49 \end{bmatrix},$$

$$\left\{ \mathbf{Z} - \hat{\mathbf{Z}}^* \right\} = \begin{bmatrix} -2 + j96 & 32 - j63 & 55 + j36 \\ 52 - j56 & -47 - j25 & 24 + j37 \\ 55 + j83 & 32 + j91 & -1 - j7 \end{bmatrix},$$

$$\left[\frac{1}{\sqrt{\hat{R}}} \right] = \begin{bmatrix} 0.125 & 0 & 0 \\ 0 & 0.111 & 0 \\ 0 & 0 & 0.200 \end{bmatrix}.$$

Multiplication of the above matrices finally gives

$$\mathbf{S}(\gamma_1 .. \gamma_j .. \gamma_N) = \begin{bmatrix} 0.518 + j0.416 & 0.086 - j0.692 & 0.168 + j0.088 \\ 0.207 - j0.551 & 0.427 + j0.227 & -0.166 + j0.370 \\ 0.091 + j0.180 & -0.375 + j0.706 & 0.526 - j0.225 \end{bmatrix}.$$

For finding $\mathbf{Z}$ from $\mathbf{S}(\gamma_1 .. \gamma_j .. \gamma_N)$, we use (7.2.6) with different matrix terms given as

$$\left[\frac{\hat{Z}}{\sqrt{R}} \right] = \begin{bmatrix} 8.00 + j3.12 & 0 & 0 \\ 0 & 9.00 - j5.56 & 0 \\ 0 & 0 & 5.00 + j15.0 \end{bmatrix},$$

$$\left\{ \left[\frac{\hat{Z}^*}{\sqrt{R}} \right] + \mathbf{S}(\gamma_1 .. \gamma_j .. \gamma_N) \left[\frac{\hat{Z}}{\sqrt{R}} \right] \right\} = \begin{bmatrix} 10.8 + j1.82 & -3.07 - j6.71 & -.476 + j2.96 \\ 3.38 - j3.76 & 14.1 + j5.22 & -6.39 - j.642 \\ .162 + j1.73 & .542 + j8.44 & 11.0 - j8.24 \end{bmatrix},$$

$$\left\{ \mathbf{U} - \mathbf{S}(\gamma_1 .. \gamma_j .. \gamma_N) \right\}^{-1} = \begin{bmatrix} 0.98 + j0.75 & 0.22 - j0.44 & 0.69 + j0.45 \\ 0.36 - j0.39 & 0.71 - j0.15 & 0.27 + j0.41 \\ 0.69 + j1.04 & 0.36 + j1.01 & 1.18 - j0.14 \end{bmatrix}$$

and finally after matrix multiplication

$$\mathbf{Z} = \begin{bmatrix} 62 + j71 & 32 - j63 & 55 + j36 \\ 52 - j56 & 34 + j25 & 24 + j37 \\ 55 + j83 & 32 + j91 & 34 - j82 \end{bmatrix}$$

which is identical to the original impedance matrix.

7.3 *N*-port Scattering Matrix with Resistive Reference-Conversion to the Generalized Scattering Matrices

The scattering parameters of a network can be measured in a system consisting of generators with resistive impedances, resistive load terminations and matched lossless guiding structures. However, it is usually convenient to express the scattering parameters of a network in terms of the references equal to the actual source and load impedance terminations of each port. It is necessary therefore, to find the relation between these latter parameters and the measured parameters. As the Z-matrix or Y-matrix of a network are independent of the source and load terminations, we can achieve the conversions by the use of these matrix representations. Thus, we can convert the scattering matrix in a system of resistive impedances to impedance or admittance matrices. The latter matrices can then be converted to the scattering matrices in any other reference system.

The above procedure is equivalent to eliminating $\mathbf{Z}$ in (7.2.5) and (7.2.6) or $\mathbf{Y}$ in (7.2.7) and (7.2.8) provided that we set $\hat{Z}_j = R_{0j}$ ($j = 1$ to N) in (7.2.6) or (7.2.8) as appropriate.

As before, we denote the S-matrices in a system of resistive references used for measurement as $\mathbf{S}$ and for a general system of references as $\mathbf{S}(\gamma_1 .. \gamma_j .. \gamma_N)$ with γ_j as previously defined.

Equation (7.2.6) with $\hat{Z}_j = R_{0j}$ ($j = 1$ to N) reduces to

$$\mathbf{Z} = [R_0]\{\mathbf{U} - \mathbf{S}\}^{-1}\{\mathbf{U} + \mathbf{S}\} \tag{7.3.1}$$

giving $\mathbf{Z}$ in terms of $\mathbf{S}$. In (7.3.1) $[R_0]$ is a diagonal matrix of jth diagonal element R_{0j}.

Equation (7.2.5), giving $\mathbf{S}(\gamma_1 .. \gamma_j .. \gamma_N)$ in terms of $\mathbf{Z}$, can be somewhat modified by defining a matrix $\boldsymbol{\gamma}$ as a diagonal matrix of elements γ_j ($j = 1$ to N) . With this definition, the diagonal reference impedance matrix $\hat{\mathbf{Z}}$ with elements

$$\hat{Z}_j = \frac{1 + \gamma_j}{1 - \gamma_j} R_{0j} \tag{7.3.2}$$

can be written as

$$\hat{\mathbf{Z}} = [R_0]\{\mathbf{U} + \boldsymbol{\gamma}\}\{\mathbf{U} - \boldsymbol{\gamma}\}^{-1} \tag{7.3.3}$$

and also

$$\hat{\mathbf{Z}}^* = [R_0]\{\mathbf{U} - \boldsymbol{\gamma}^*\}^{-1}\{\mathbf{U} + \boldsymbol{\gamma}^*\}. \tag{7.3.4}$$

We also define a matrix $\boldsymbol{\Lambda}$ as a diagonal matrix of elements Λ_j with

$$\Lambda_j = (1 - \gamma_j^*)\sqrt{\frac{1 - \gamma_j \gamma_j^*}{(1 - \gamma_j)(1 - \gamma_j^*)}}. \tag{7.3.5}$$

With this definition it is clear that

$$\Lambda_j \Lambda_j{}^* = 1 - \gamma_j \gamma_j{}^*.$$

From the above relations, the jth element of the diagonal matrix $\sqrt{\hat{R}_j}$ can be written as

$$\sqrt{\hat{R}_j} = \sqrt{\frac{\hat{Z}_j + \hat{Z}_j{}^*}{2}} = \sqrt{R_{0j}} \; \sqrt{\frac{1}{2}\left[\frac{1+\gamma_j}{1-\gamma_j} + \frac{1+\gamma_j{}^*}{1-\gamma_j{}^*}\right]} = \sqrt{R_{0j}} \; \frac{\Lambda_j{}^*}{1-\gamma_j}.$$

Hence in matrix form,

$$\left[\sqrt{\hat{R}}\right] = \left[\sqrt{R_0}\right]\{U - \gamma\}^{-1}\Lambda^*, \tag{7.3.6}$$

$$\left[\frac{1}{\sqrt{\hat{R}}}\right] = \left[\frac{1}{\sqrt{R_0}}\right]\Lambda^{-1}\{U - \gamma^*\}. \tag{7.3.7}$$

Substituting into (7.2.5) and making some rearrangements as shown in the Appendix, we finally obtain

$$S(\gamma_1 .. \gamma_j .. \gamma_N) = \Lambda^{-1}\{S - \gamma^*\}\{U - \gamma S\}^{-1}\Lambda^*. \tag{7.3.8}$$

The generalized scattering parameters $S(\gamma_1 .. \gamma_j .. \gamma_N)$ can also be converted to S by some rearrangement of (7.3.8) giving

$$S = \left\{\Lambda S(\gamma_1 .. \gamma_j .. \gamma_N)\Lambda^{*-1}\gamma + U\right\}^{-1}\left\{\Lambda S(\gamma_1 .. \gamma_j .. \gamma_N)\Lambda^{*-1} + \gamma^*\right\}.$$

Finding S as an intermediate step, the scattering parameters of a network for any set of reference impedances can readily be converted to the generalized scattering matrix of any other set.

Example (7.3.1)

From expression (7.3.8) for the conversion of the scattering parameters in a system of Z_j ($j = 1$ to N) = R_{0j} to a general system, deduce the corresponding expression (6.3.3) for a 2-port network.

Solution

Equation (7.3.8) applied to a 2-port network can be written as

$$
\begin{bmatrix} S_{11}(\gamma_1,\gamma_2) & S_{12}(\gamma_1,\gamma_2) \\ S_{21}(\gamma_1,\gamma_2) & S_{22}(\gamma_1,\gamma_2) \end{bmatrix}
$$

$$
= \begin{bmatrix} \Lambda_1^{-1} & 0 \\ 0 & \Lambda_2^{-1} \end{bmatrix} \begin{bmatrix} S_{11}-\gamma_1^* & S_{12} \\ S_{21} & S_{22}-\gamma_2^* \end{bmatrix} \begin{bmatrix} 1-\gamma_1 S_{11} & -\gamma_1 S_{12} \\ -\gamma_2 S_{21} & 1-\gamma_2 S_{22} \end{bmatrix}^{-1} \begin{bmatrix} \Lambda_1^* & 0 \\ 0 & \Lambda_2^* \end{bmatrix}
$$

$$
= \frac{1}{\Sigma_S} \begin{bmatrix} \Lambda_1^{-1} & 0 \\ 0 & \Lambda_2^{-1} \end{bmatrix} \begin{bmatrix} S_{11}-\gamma_1^* & S_{12} \\ S_{21} & S_{22}-\gamma_2 \end{bmatrix} \begin{bmatrix} 1-\gamma_2 S_{22} & \gamma_1 S_{12} \\ \gamma_2 S_{21} & 1-\gamma_1 S_{11} \end{bmatrix} \begin{bmatrix} \Lambda_1^* & 0 \\ 0 & \Lambda_2^* \end{bmatrix}
$$

$$
= \frac{1}{\Sigma_S} \begin{bmatrix} \dfrac{S_{11}-\gamma_1^*}{\Lambda_1} & \dfrac{S_{12}}{\Lambda_1} \\ \dfrac{S_{21}}{\Lambda_2} & \dfrac{S_{22}-\gamma_2^*}{\Lambda_2} \end{bmatrix} \begin{bmatrix} \Lambda_1^*(1-\gamma_2 S_{22}) & \Lambda_2^*\gamma_1 S_{12} \\ \Lambda_1^*\gamma_2 S_{21} & \Lambda_2^*(1-\gamma_1 S_{11}) \end{bmatrix}
$$

where $\Sigma_S = (1-\gamma_1 S_{11})(1-\gamma_2 S_{22}) - \gamma_1\gamma_2 S_{12}S_{21}$.

After the required matrix multiplication, Equation (6.3.3) can easily be found.

Example (7.3.2)

Convert the S-matrix of a 3-port network in a system of $\hat{Z}_1 = R_{01}$, $\hat{Z}_2 = R_{02}$ and $\hat{Z}_3 = R_{03}$ given as

$$
S = \begin{bmatrix} S_{11} & S_{12} & S_{13} \\ S_{21} & S_{22} & S_{23} \\ S_{31} & S_{32} & S_{33} \end{bmatrix}
$$

to the S-matrix in a general system.

Solution

For the 3-port network we can write

$$\{S - \gamma^*\} = \begin{bmatrix} (S_{11} - \gamma_1^*) & S_{12} & S_{13} \\ S_{21} & (S_{22} - \gamma_2^*) & S_{23} \\ S_{31} & S_{32} & (S_{33} - \gamma_3^*) \end{bmatrix}$$

and

$$\{U - \gamma\,S\}^{-1} = \frac{1}{D} \times$$

$$\begin{bmatrix} 1 - \gamma_2 S_{22} - \gamma_3 S_{33} + \gamma_2\gamma_3\Delta_{11} & \gamma_1 S_{12} + \gamma_1\gamma_3\Delta_{21} & \gamma_1 S_{13} + \gamma_1\gamma_2\Delta_{31} \\ \gamma_2 S_{21} + \gamma_2\gamma_3\Delta_{12} & 1 - \gamma_1 S_{11} - \gamma_3 S_{33} + \gamma_1\gamma_3\Delta_{22} & \gamma_2 S_{23} + \gamma_1\gamma_2\Delta_{32} \\ \gamma_3 S_{31} + \gamma_2\gamma_3\Delta_{13} & \gamma_3 S_{32} + \gamma_1\gamma_3\Delta_{23} & 1 - \gamma_1 S_{11} - \gamma_2 S_{22} + \gamma_1\gamma_2\Delta_{33} \end{bmatrix}$$

where D is the determinant of the above matrix and is given by

$$D = 1 - \gamma_1 S_{11} - \gamma_2 S_{22} - \gamma_3 S_{33} + \gamma_1\gamma_2\Delta_{33} + \gamma_1\gamma_3\Delta_{22} + \gamma_2\gamma_3\Delta_{11} - \gamma_1\gamma_2\gamma_3\Delta$$

with Δ_{jk} as the cofactor of S_{jk} and Δ the determinant of the original scattering matrix given as

$$\Delta = S_{j1}\Delta_{j1} + S_{j2}\Delta_{j2} + S_{j3}\Delta_{j3} \quad (j = 1,2,3).$$

After substitution into (7.3.8) and expansion we finally find

$$S_{11}(\gamma_1, \gamma_2, \gamma_3) = \frac{\Lambda_1^*}{\Lambda_1 D}[(S_{11} - \gamma_2\Delta_{33} - \gamma_3\Delta_{22} + \gamma_2\gamma_3\Delta) - \gamma_1^*(1 - \gamma_2 S_{22} - \gamma_3 S_{33} + \gamma_2\gamma_3\Delta_{11})],$$

$$S_{12}(\gamma_1, \gamma_2, \gamma_3) = \frac{1}{D}\Lambda_2^*\Lambda_1^* S_{12}(1 + \gamma_3\Delta_{21} / S_{12}),$$

$$S_{13}(\gamma_1, \gamma_2, \gamma_3) = \frac{1}{D}\Lambda_3^*\Lambda_1^* S_{13}(1 + \gamma_2\Delta_{31} / S_{13}),$$

$$S_{21}(\gamma_1,\gamma_2,\gamma_3) = \frac{1}{D}\Lambda_1^*\Lambda_2^* S_{21}(1+\gamma_3\Delta_{12}/S_{21}),$$

$$S_{22}(\gamma_1,\gamma_2,\gamma_3) = \frac{\Lambda_2^*}{\Lambda_2 D}[(S_{22}-\gamma_1\Delta_{33}-\gamma_3\Delta_{11}+\gamma_1\gamma_3\Delta)-\gamma_2^*(1-\gamma_1 S_{11}-\gamma_3 S_{33}+\gamma_1\gamma_3\Delta_{22})],$$

$$S_{23}(\gamma_1,\gamma_2,\gamma_3) = \frac{1}{D}\Lambda_3^*\Lambda_2^* S_{23}(1+\gamma_1\Delta_{32}/S_{23}),$$

$$S_{31}(\gamma_1,\gamma_2,\gamma_3) = \frac{1}{D}\Lambda_1^*\Lambda_3^* S_{31}(1+\gamma_2\Delta_{13}/S_{31}),$$

$$S_{32}(\gamma_1,\gamma_2,\gamma_3) = \frac{1}{D}\Lambda_2^*\Lambda_3^* S_{32}(1+\gamma_1\Delta_{23}/S_{32}),$$

$$S_{33}(\gamma_1,\gamma_2,\gamma_3) = \frac{\Lambda_3^*}{\Lambda_3 D}[(S_{33}-\gamma_1\Delta_{22}-\gamma_2\Delta_{11}+\gamma_1\gamma_2\Delta)-\gamma_3^*(1-\gamma_1 S_{11}-\gamma_2 S_{22}+\gamma_1\gamma_2\Delta_{33})].$$

Example (7.3.3)

The measured scattering matrix $(\hat{Z}_1 = \hat{Z}_2 = \hat{Z}_3 = 50)$ of a 3-port network is given by

$$S = \begin{bmatrix} 0.648+j0.297 & -0.073-j0.449 & 0.034+j0.150 \\ 0.011-j0.422 & 0.541+j0.359 & -0.134+j0.241 \\ -0.119+j0.186 & -0.228+j0.456 & 0.645-j0.416 \end{bmatrix}$$

Find the scattering matrix in a system of $\hat{Z}_1 = 64+j25$, $\hat{Z}_2 = 81-j50$ and $\hat{Z}_3 = 25+j75$.

Solution

Different matrices in expression (7.3.8) can be written as

$$\gamma = \begin{bmatrix} 0.163+j0.184 & 0 & 0 \\ 0 & 0.334-j0.254 & 0 \\ 0 & 0 & 0.333+j0.667 \end{bmatrix},$$

$$\{U - \gamma S\} = \begin{bmatrix} 0.949 - j0.167 & -0.071 + j0.087 & 0.022 - j0.031 \\ 0.104 + j0.144 & 0.728 + j0.018 & -0.017 - j0.115 \\ 0.164 + j0.017 & 0.380 + j0.000 & 0.508 - j0.291 \end{bmatrix},$$

$$\{U - \gamma S\}^{-1} = \begin{bmatrix} 0.999 + j0.162 & 0.138 - j0.134 & -0.048 + j0.052 \\ -0.118 - j0.252 & 1.354 - j0.186 & -0.051 + j0.274 \\ -0.216 - j0.022 & -0.876 - j0.325 & 1.619 + j0.709 \end{bmatrix},$$

$$\{S - \gamma^*\} = \begin{bmatrix} 0.485 + j0.481 & -0.073 - j0.449 & 0.034 + j0.150 \\ 0.011 - j0.422 & 0.207 + j0.105 & -0.134 + j0.241 \\ -0.119 + j0.186 & -0.228 + j0.456 & 0.312 + j0.251 \end{bmatrix},$$

$$\Lambda = \begin{bmatrix} 0.947 + j0.208 & 0 & 0 \\ 0 & 0.848 - j0.324 & 0 \\ 0 & 0 & 0.471 + j0.471 \end{bmatrix}.$$

And finally after matrix multiplication

$$S(\gamma_1, \gamma_2, \gamma_3) = \begin{bmatrix} 0.521 + j0.417 & 0.072 - j0.685 & 0.167 + j0.085 \\ 0.206 - j0.545 & 0.411 + j0.201 & -0.161 + j0.368 \\ 0.080 + j0.170 & -0.294 + j0.699 & 0.526 - j0.209 \end{bmatrix}.$$

Alternatively we could use the expressions of the previous example, which would lead to the same results.

7.4 The Wave Amplitudes of a General N-port Network

In this section we derive the wave amplitudes of an N-port network connected to a required set of generators and load impedances. In general each port of an N-port network may be connected to a generator. Assuming the system to be linear, however, we can consider the total contribution to wave amplitudes at any port as the summations of the contributions from each generator with all other generators replaced by their generator impedances. The contribution from each source can hence by considered separately.

To simplify the problem, therefore, we consider an N-port network connected at its port 1 to a source of emf E_S and impedance Z_{S1}. We designate the impedance of the remaining $(N - 1)$ ports as $Z_{L2},..~Z_{Lj},..$ and Z_{LN}. The designation of port 1 to the port with a generating source is for convenience.

The incident and reflected wave amplitudes at each port are not unique, but depend on the chosen reference impedances. It is convenient to consider either: (a) a system of reference impedances given as $\hat{Z}_1 = Z_{S1}$, $\hat{Z}_j = Z_{Lj}$ (j = 2 to N) or (b) a system of reference impedances given as $\hat{Z}_j = R_{0j}$ (j = 1 to N), where R_{0j} is the source and load impedances, as well as, the characteristic impedance of the guiding structure for the jth port. We consider these two situations separately.

(a) Analysis in a system of references impedances $\hat{Z}_1 = Z_{S1}$, $\hat{Z}_j = Z_{Lj}$ *(j = 2 to N)*

In this case the derivation of the wave amplitudes is very simple. Apart from port 1 that is terminated by a generating source and the source impedance, all other ports are terminated by their reference impedances. Hence we have

$$a_1(\Gamma_1..\Gamma_j..\Gamma_N) = e_S, \qquad (7.4.1)$$

$$a_j(\Gamma_1..\Gamma_j..\Gamma_N) = 0 \qquad (7.4.2)$$

where

$$e_S = \frac{E_S}{2\sqrt{R_{S1}}},$$

$$\Gamma_1 = \frac{\hat{Z}_1 - R_{0j}}{\hat{Z}_1 + R_{0j}} = \frac{Z_{S1} - R_{0j}}{Z_{S1} + R_{0j}},$$

and

$$\Gamma_j = \frac{\hat{Z}_j - R_{0j}}{\hat{Z}_j + R_{0j}} = \frac{Z_{Lj} - R_{0j}}{Z_{Lj} + R_{0j}} \qquad (j = 2 \text{ to } N).$$

The schematic representation of the network is given in Fig. (7.4.1). The matrix relation

$$\mathbf{b}(\Gamma_1..\Gamma_j..\Gamma_N) = \mathbf{S}(\Gamma_1..\Gamma_j..\Gamma_N)\mathbf{a}(\Gamma_1..\Gamma_j..\Gamma_N) \qquad (7.4.3)$$

provides the required relations for the remaining wave amplitudes. For the jth term

$$b_j(\Gamma_1..\Gamma_j..\Gamma_N) = e_S S_{j1}(\Gamma_1..\Gamma_j..\Gamma_N) \qquad j = 1 \text{ to } N. \qquad (7.4.4)$$

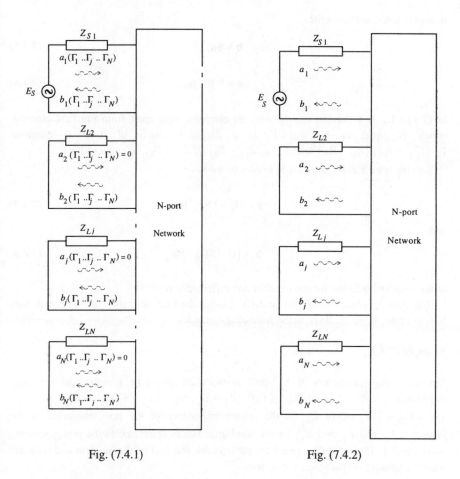

Fig. (7.4.1)   Fig. (7.4.2)

(b) Analysis in a system of references $\hat{Z}_j = R_{0j}$ $(j = 1$ to $N)$

The schematic representation of the network is given in Fig. (7.4.2). N linear equations relate the reflected parameters b_j to the incident parameters a_j by $N \times N$ scattering parameters. In addition we have $a_1 = \Gamma_S b_1 + b_S$ and $a_j = \Gamma_j b_j$ $(j = 2$ to $N)$, where

$$b_S = \frac{E_S}{2\sqrt{R_{01}}}(1 - \Gamma_S)$$

providing the additional N required equations for the $2N$ unknowns.

In matrix form, we can write

$$\mathbf{b} = \mathbf{Sa},\qquad(7.4.5)$$

$$\mathbf{a} = \mathbf{\Gamma}\,\mathbf{b} + \mathbf{b}_S.\qquad(7.4.6)$$

In (7.4.6) $\mathbf{b}_S$ is a column matrix with all elements zero apart from the first element, which is equal to b_S and $\mathbf{\Gamma}$ is a diagonal matrix of diagonal elements $\Gamma_1 = (Z_{S1} - R_{01})/(Z_{S1} + R_{01})$ and $\Gamma_j = (Z_{Lj} - R_{0j})/(Z_{Lj} + R_{0j})$ ($j = 2$ to N).
Solving for $\mathbf{a}$ and $\mathbf{b}$ from (7.4.5) to (7.4.6) we have

$$\mathbf{a} = \{\mathbf{U} - \mathbf{\Gamma S}\}^{-1}\mathbf{b}_S\qquad(7.4.7)$$

and

$$\mathbf{b} = \{\mathbf{U} - \mathbf{S\Gamma}\}^{-1}\mathbf{S}\mathbf{b}_S\qquad(7.4.8)$$

as the required solution for the incident and reflected waves.

The wave amplitudes can also be derived using the flow diagrams in connection with Mason's rule, as we shall see in the following examples.

Example (7.4.1)

The scattering parameters of a 3-port network are given in a system of reference impedances $\hat{Z}_1 = Z_{S1}(\gamma_1 = \Gamma_{S1} = \Gamma_1)$, $\hat{Z}_2 = Z_{L2}(\gamma_2 = \Gamma_{L2} = \Gamma_2)$ and $\hat{Z}_3 = Z_{L3}$ ($\gamma_3 = \Gamma_{L3} = \Gamma_3$), where Z_{S1} is the source impedance of the port connected to the generator E_S and Z_{L2} and Z_{L3} are the load impedances connected to the non-generating ports 2 and 3. Draw the flow graph for the network and find all the incident and reflected wave amplitudes in that reference system.

Solution

The flow graph representation is as shown. The incident and reflected wave amplitudes can be written readily as

$$a_1(\Gamma_{S1}, \Gamma_{L2}, \Gamma_{L3}) = e_S,$$

$$a_2(\Gamma_{S1}, \Gamma_{L2}, \Gamma_{L3}) = 0,$$

$$a_3(\Gamma_{S1}, \Gamma_{L2}, \Gamma_{L3}) = 0,$$

$$b_1(\Gamma_{S1}, \Gamma_{L2}, \Gamma_{L3}) = e_S S_{11}(\Gamma_{S1}, \Gamma_{L2}, \Gamma_{L3}),$$

$$b_2(\Gamma_{S1}, \Gamma_{L2}, \Gamma_{L3}) = e_S S_{21}(\Gamma_{S1}, \Gamma_{L2}, \Gamma_{L3}),$$

$$b_3(\Gamma_{S1}, \Gamma_{L2}, \Gamma_{L3}) = e_S S_{31}(\Gamma_{S1}, \Gamma_{L2}, \Gamma_{L3})$$

where $e_S = \dfrac{E_S}{2\sqrt{R_{S1}}}$ with $R_{S1} = \text{Re}(Z_{S1})$.

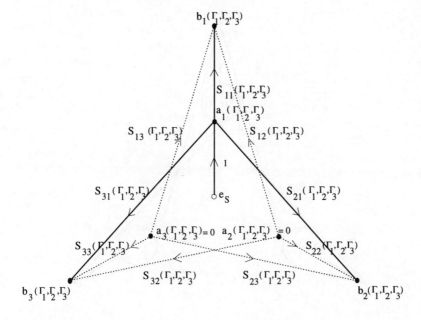

Example (7.4.2)

The scattering parameters of the 3-port network of the previous example are given in a system of reference impedances $\hat{Z}_1 = R_{01}$, $\hat{Z}_2 = R_{02}$ and $\hat{Z}_3 = R_{03}$. Draw the flow graph for the network and find the wave amplitudes.

Solution

The flow graph for this reference impedance system is shown. To find the node values we identify the required loops and paths as defined in Section 4 of Chapter 2, as well as, the non-touching loops for each path as given below.

(i) 1st order loops:

$\Gamma_1 S_{11}, \quad \Gamma_2 S_{22}, \quad \Gamma_3 S_{33}$

$\Gamma_1 S_{21} \Gamma_2 S_{12}, \quad \Gamma_2 S_{32} \Gamma_3 S_{23}, \quad \Gamma_3 S_{13} \Gamma_1 S_{31}$

$\Gamma_1 S_{21} \Gamma_2 S_{32} \Gamma_3 S_{13}, \quad \Gamma_2 S_{12} \Gamma_3 S_{23} \Gamma_1 S_{31}$

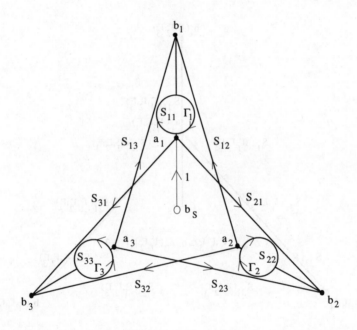

(ii) 2nd order loops:

$\Gamma_1 S_{11} \Gamma_2 S_{22}, \quad \Gamma_2 S_{22} \Gamma_3 S_{33}, \quad \Gamma_3 S_{33} \Gamma_1 S_{11}$

$\Gamma_1 S_{11} \Gamma_2 S_{32} \Gamma_3 S_{23}, \quad \Gamma_2 S_{22} \Gamma_3 S_{13} \Gamma_1 S_{31}, \quad \Gamma_3 S_{33} \Gamma_1 S_{21} \Gamma_2 S_{12}$

(iii) 3rd order loops:

$\Gamma_1 S_{11} \Gamma_2 S_{22} \Gamma_3 S_{33}$

(iv) Paths from b_S to b_1

Path 1: $(1) \times S_{11}$: non-touching loops;

$\quad$ 1st order; $\Gamma_2 S_{22}$, $\Gamma_3 \Gamma_{33}$, $\Gamma_2 S_{32} \Gamma_3 S_{23}$

$\quad$ 2nd order: $\Gamma_2 S_{22} \Gamma_3 S_{33}$

Path 2: $(1) \times S_{21} \Gamma_2 S_{12}$: non-touching loops;

$\quad$ 1st order; $\Gamma_3 S_{33}$

Path 3: $(1) \times S_{31} \Gamma_3 S_{13}$: non-touching loops;

$\quad$ 1st order; $\Gamma_2 S_{22}$

Path 4: $(1) \times S_{21} \Gamma_2 S_{32} \Gamma_3 S_{13}$: non-touching loops; none

Path 5: $(1) \times S_{31} \Gamma_3 S_{23} \Gamma_2 S_{12}$: non-touching loops; none

(v) Paths from b_S to b_2

Path 1: $(1) \times S_{21}$: non-touching loops;

$\quad$ 1st order; $\Gamma_3 S_{33}$

Path 2: $(1) \times S_{31} \Gamma_3 S_{23}$: non-touching loop; none

(vi) Paths from b_S to b_3

Path 1: $(1) \times S_{31}$: non-touching loops;

$\quad$ 1st order; $\Gamma_S S_{22}$

Path 2: $(1) \times S_{21} \Gamma_2 S_{32}$: non-touching loop; none

Hence by the application of Mason's Rule, we have

$$b_1 = b_S[S_{11}(1 - \Gamma_2 S_{22} - \Gamma_3 S_{33} - \Gamma_2 S_{32}\Gamma_3 S_{23} + \Gamma_2 S_{22}\Gamma_3 S_{33}]$$

$$+ S_{21}\Gamma_2 S_{21}(1 - \Gamma_3 S_{33}) + S_{31}\Gamma_3 S_{13}(1 - \Gamma_2 S_{22})$$

$$+ S_{21}\Gamma_2 S_{32}\,\Gamma_3 S_{13} + S_{31}\Gamma_3 S_{23}\Gamma_2 S_{12}] / D,$$

$$b_2 = b_S[S_{21}(1 - \Gamma_3 S_{33}) + S_{31}\Gamma_3 S_{23}] / D,$$

$$b_3 = b_S[S_{31}(1 - \Gamma_2 S_{22}) + S_{21}\Gamma_2 S_{32}] / D$$

where $b_S = E_S(1 - \Gamma_1)/2\sqrt{R_{01}}$ and

$$D = 1 - \Gamma_1 S_{11} - \Gamma_2 S_{22} - \Gamma_3 S_{33}$$

$$- \Gamma_1 S_{21}\Gamma_2 S_{12} - \Gamma_2 S_{32}\Gamma_3 S_{23} - \Gamma_3 S_{13}\Gamma_1 S_{31}$$

$$- \Gamma_1 S_{21}\Gamma_2 S_{32}\Gamma_3 S_{13} - \Gamma_2 S_{12}\Gamma_3 S_{23}\Gamma_1 S_{31}$$

$$+ \Gamma_1 S_{11}\Gamma_2 S_{22} + \Gamma_2 S_{22}\Gamma_3 S_{33} + \Gamma_1 S_{11}\Gamma_3 S_{33}$$

$$+ \Gamma_1 S_{11}\Gamma_2 S_{32}\Gamma_3 S_{23} + \Gamma_2 S_{22}\Gamma_3 S_{13}\Gamma_1 S_{31} + \Gamma_3 S_{33}\Gamma_1 S_{21}\Gamma_2 S_{12}$$

$$- \Gamma_1 S_{11}\Gamma_2 S_{22}\Gamma_3 S_{33}$$

$$= 1 - \Gamma_1 S_{11} - \Gamma_2 S_{22} - \Gamma_3 S_{33} + \Gamma_1\Gamma_2\Delta_{33} + \Gamma_1\Gamma_3\Delta_{22} + \Gamma_2\Gamma_3\Delta_{11} - \Gamma_1\Gamma_2\Gamma_3\Delta$$

as in Example (7.3.2).

The above expressions can also be verified using Equation (7.4.8). In this equation

$$\mathbf{S}b_S = b_S\begin{bmatrix} S_{11} \\ S_{21} \\ S_{31} \end{bmatrix}, \quad \{\mathbf{U} - \mathbf{S}\Gamma\} = \begin{bmatrix} (1 - \Gamma_1 S_{11}) & -\Gamma_2 S_{12} & -\Gamma_3 S_{13} \\ \Gamma_1 S_{21} & (1 - \Gamma_2 S_{22}) & -\Gamma_3 S_{23} \\ -\Gamma_1 S_{31} & -\Gamma_2 S_{32} & (1 - \Gamma_3 S_{33}) \end{bmatrix}$$

or

$$\{U - S\Gamma\}^{-1} = \frac{1}{D} \times$$

$$
\begin{bmatrix}
\begin{array}{l}(1 - \Gamma_2 S_{22})(1 - \Gamma_3 S_{33}) \\ -\Gamma_2 S_{32} \Gamma_3 S_{23}\end{array} & \begin{array}{l}\Gamma_2 S_{12}(1 - \Gamma_3 S_{33}) \\ +\Gamma_2 \Gamma_3 S_{13} S_{32}\end{array} & \begin{array}{l}\Gamma_2 \Gamma_3 S_{12} S_{23} \\ +\Gamma_3 S_{13}(1 - \Gamma_2 S_{22})\end{array} \\[2ex]
\begin{array}{l}\Gamma_1 S_{21}(1 - \Gamma_3 S_{33}) \\ +\Gamma_1 \Gamma_3 S_{23} S_{31}\end{array} & \begin{array}{l}(1 - \Gamma_1 S_{11})(1 - \Gamma_3 S_{33}) \\ -\Gamma_1 S_{31} \Gamma_3 S_{13}\end{array} & \begin{array}{l}\Gamma_3 S_{23}(1 - \Gamma_1 S_{11}) \\ +\Gamma_1 \Gamma_3 S_{13} S_{21}\end{array} \\[2ex]
\begin{array}{l}\Gamma_1 S_{21} \Gamma_2 S_{32} \\ +\Gamma_1 S_{31}(1 - \Gamma_2 S_{22})\end{array} & \begin{array}{l}\Gamma_2 S_{32}(1 - \Gamma_1 S_{11}) \\ +\Gamma_1 S_{31} \Gamma_2 S_{12}\end{array} & \begin{array}{l}(1 - \Gamma_1 S_{11})(1 - \Gamma_2 S_{22}) \\ -\Gamma_1 S_{21} \Gamma_2 S_{12}\end{array}
\end{bmatrix}
$$

which after the required matrix multiplication leads to the results previously found.

7.5 Reduction to Two Ports and Possible Reverse Process

In many applications of the N-port networks, any two particular ports of the network can be considered as the network input and output ports, while all other ports can be termed as the control ports of the network. The scattering matrix for an N-port network can then be reduced to an equivalent 2-port network scattering matrix, relating the reflected wave amplitudes of the input and output ports to that of the incident wave amplitudes. In this section we consider the N-port network reduction to an equivalent 2-port network. The 2-port network scattering parameters calculated dependent however, in addition to the scattering parameters of the original network, on the control port terminations. Thus the properties of the reduced 2-port network such as the stability and gain will depend also on the terminations of the control ports.

For convenience we designate the input port of the network as port 1 and the output port as port 2. Reduction to two ports is particularly simple in a reference system with $\hat{Z}_1 = Z_S$, $\hat{Z}_2 = Z_L$ and $\hat{Z}_j = Z_{Lj}$ ($j = 3$ to N), where Z_S and Z_L are the source and load impedances of port 1 and 2 respectively and Z_{Lj} the terminating impedance of the jth port ($j \neq 1$, $j \neq 2$). The N-port scattering matrix can be denoted as

$$\mathbf{S}(\Gamma_S, \Gamma_L, ... \Gamma_j, ... \Gamma_N)$$

where

$$\Gamma_j = \frac{Z_{Lj} - R_{0j}}{Z_{Lj} + R_{0J}} \quad (j = 3 \text{ to } N)$$

and Γ_S and Γ_L, similarly related to Z_S and Z_L.

In this system of references $S_{11}(\Gamma_S,\Gamma_L,..\Gamma_j,..\Gamma_N)$, $S_{12}(\Gamma_S,\Gamma_L,..\Gamma_j,..\Gamma_N)$, $S_{21}(\Gamma_S,\Gamma_L,..\Gamma_j,..\Gamma_N)$ and $S_{22}(\Gamma_S,\Gamma_L,..\Gamma_j,..\Gamma_N)$ are simply the required equivalent 2-port scattering parameters.

The above scattering parameters can be expressed in terms of the scattering parameters $S_{11}(0,0,..\Gamma_j..\Gamma_N)$, $S_{12}(0,0,..\Gamma_j..\Gamma_N)$, $S_{21}(0,0,..\Gamma_j..\Gamma_N)$ and $S_{22}(0,0,..\Gamma_j..\Gamma_N)$, by substituting these parameters in (6.3.3) for S_{11}, S_{12}, S_{21} and S_{22} respectively and setting $\gamma_1 = \Gamma_S$ and $\gamma_2 = \Gamma_L$. The scattering parameters $S_{11}(0,0,..\Gamma_j..\Gamma_N)$, $S_{12}(0,0,..\Gamma_j..\Gamma_N)$, $S_{21}(0,0,..\Gamma_j..\Gamma_N)$ and $S_{22}(0,0,..\Gamma_j..\Gamma_N)$ are the scattering parameters that can be measured, with control ports appropriately terminated.

The reverse process - Indefinite scattering matrices

In general the knowledge of the four scattering parameters of the equivalent 2-port network is not sufficient for derivation of $N \times N$ scattering parameters of an N-port network. In fact, unless we have some other constraints, $(N^2 - 4)$ parameters can be given arbitrary values. We may be able, however, to find a unique set of scattering parameters, if additional constraints or properties of the network are specified.

As an example consider a 3-port network with four given scattering parameters and the following constraints

$$\sum_{j=1}^{3} S_{kj} = 1 \qquad (k = 1 \text{ to } 3) \tag{7.5.1}$$

and

$$\sum_{k=1}^{3} S_{kj} = 1 \qquad (j = 1 \text{ to } 3) \tag{7.5.2}$$

In this example, we assume that the scattering parameters are given in a system of references $\hat{Z}_1 = \hat{Z}_2 = \hat{Z}_3 = R_0$. Results of Section (7.3) can be used for conversion to any other system.

For the above network, (7.5.1) and (7.5.2) provide the five independent equations, which are required for determination of 5 unknown parameters. Such a network is represented in Fig. (7.5.1a). A general 3-port network is given in Fig. (7.5.1b) for comparison.

The difference between the network in Fig. (7.5.1a) and Fig. (7.5.1b) is that in the case of (a) the return terminals of the ports, which are common and float, have no direct connections to the rest of the network. Consequently the total sum of the incident and reflected currents for three ports is equal to zero. With the wave amplitudes proportional to currents, we have

$$a_1 + a_2 + a_3 - b_1 - b_2 - b_3 = 0. \tag{7.5.3}$$

The above relation between the wave amplitudes should hold for every value of a_1, a_2 and a_3. Hence

setting $a_2(E_1 = 0, \quad Z_1 = R_0) = a_3(E_3 = 0, \quad Z_3 = R_0) = 0$ we have $S_{11} + S_{21} + S_{31} = 1$,

setting $a_3(E_3 = 0, \quad Z_3 = R_0) = a_1(E_1 = 0, \quad Z_1 = R_0) = 0$ we have $S_{12} + S_{22} + S_{32} = 1$,

setting $a_1(E_1 = 0, \quad Z_1 = R_0) = a_2(E_2 = 0, \quad \hat{Z}_2 = R_0) = 0$ we have $S_{13} + S_{23} + S_{33} = 1$.

Thus (7.5.1) is satisfied.

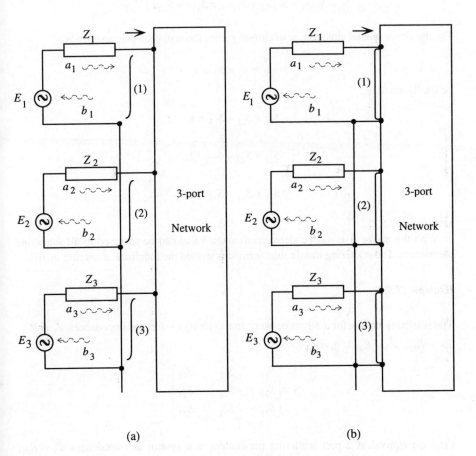

(a)   (b)

Fig. (7.5.1)

Next we let

$$a_1 = a_2 = a_3$$

by setting $E_1 = E_2 = E_3$. From the definition of the scattering parameters we can write

$$b_1 = S_{11}a_1 + S_{12}a_2 + S_{13}a_3 = S_{11}a + S_{12}a + S_{13}a,$$

$$b_2 = S_{21}a_1 + S_{22}a_2 + S_{23}a_3 = S_{21}a + S_{22}a + S_{23}a,$$

$$b_3 = S_{31}a_1 + S_{32}a_2 + S_{33}a_3 = S_{31}a + S_{32}a + S_{33}a.$$

For the above special situation, zero current passes through each port and hence

$$b_1 = b_2 = b_3 = a$$

or finally we have

$$S_{11} + S_{12} + S_{13} = 1,$$

$$S_{21} + S_{22} + S_{23} = 1,$$

$$S_{31} + S_{32} + S_{33} = 1$$

satisfying (7.5.2).

With the above five independent equations, a value can be assigned to all scattering parameters. The scattering matrix thus formed is termed the indefinite scattering matrix.

Example (7.5.1)

The scattering matrix for a 3-port network in a system of reference impedances $\hat{Z}_1 = R_{01}$, $\hat{Z}_2 = R_{02}$, $\hat{Z}_3 = R_{03}$ is given as

$$\mathbf{S} = \begin{bmatrix} S_{11} & S_{12} & S_{13} \\ S_{21} & S_{22} & S_{23} \\ S_{31} & S_{32} & S_{33} \end{bmatrix}.$$

Find the equivalent 2-port scattering parameters in a system of impedances $\hat{Z}_1 = R_{01}$, $\hat{Z}_2 = R_{02}$, if the third port is terminated by a load of impedance Z_{L3}.

Solution

With

$$\Gamma_3 = \frac{Z_{L3} - R_{03}}{Z_{L3} + R_{03}}$$

the required scattering parameters are $S_{11}(0,0,\Gamma_3)$, $S_{12}(0,0,\Gamma_3)$, $S_{21}(0,0,\Gamma_3)$ and $S_{22}(0,0,\Gamma_3)$. In Example (7.3.2) we found the generalized scattering parameters of a 3-port network in terms of measurable scattering parameters. Substituting $\Gamma_1 = 0$ and $\Gamma_2 = 0$ in these expressions, we find

$$S_{11}(0,0,\Gamma_3) = \frac{S_{11} - \Gamma_3 \Delta_{22}}{1 - \Gamma_3 S_{33}}, \quad \leftarrow$$

$$S_{12}(0,0,\Gamma_3) = \frac{S_{12} + \Gamma_3 \Delta_{21}}{1 - \Gamma_3 S_{33}}, \quad \leftarrow$$

$$S_{13}(0,0,\Gamma_3) = \frac{\Lambda_3^* S_{13}}{1 - \Gamma_3 S_{33}},$$

$$S_{21}(0,0,\Gamma_3) = \frac{S_{21} + \Gamma_3 \Delta_{12}}{1 - \Gamma_3 S_{33}}, \quad \leftarrow$$

$$S_{22}(0,0,\Gamma_3) = \frac{S_{22} - \Gamma_3 \Delta_{11}}{1 - \Gamma_3 S_{33}}, \quad \leftarrow$$

$$S_{23}(0,0,\Gamma_3) = \frac{\Lambda_3^* S_{23}}{1 - \Gamma_3 S_{33}},$$

$$S_{31}(0,0,\Gamma_3) = \frac{\Lambda_3^* S_{31}}{1 - \Gamma_3 S_{33}},$$

$$S_{32}(0,0,\Gamma_3) = \frac{\Lambda_3^* S_{32}}{1 - \Gamma_3 S_{33}},$$

$$S_{33}(0,0,\Gamma_3) = \frac{\Lambda_3^*}{\Lambda_3} \frac{S_{33} - \Gamma_3^*}{1 - \Gamma_3 S_{33}}.$$

$S_{11}(0,0,\Gamma_3)$, $S_{12}(0,0,\Gamma_3)$, $S_{21}(0,0,\Gamma_3)$ and $S_{22}(0,0,\Gamma_3)$ are the required expressions. Where Δ_{11} and other similar terms are previously defined.

Example (7.5.2)

The scattering parameters of a 3-port network with ports designated as A, B and C are given in Common C configuration (port C short circuited) as $S_{11}{}^C$, $S_{12}{}^C$, $S_{21}{}^C$ and $S_{22}{}^C$. Find the scattering parameter of the 3-port network in Common B and Common A configurations.

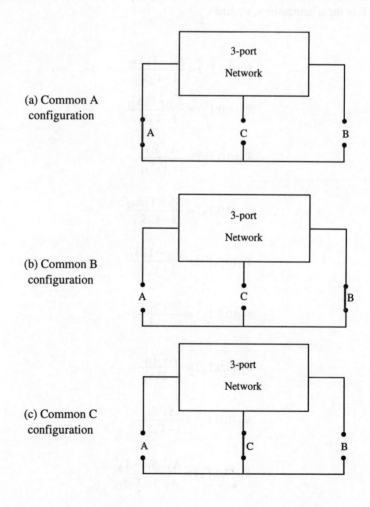

(a) Common A configuration

(b) Common B configuration

(c) Common C configuration

Solution

From the knowledge of the common C scattering parameters, we first form the indefinite scattering matrix and subsequently derive the common B and common A scattering parameters. The three possible configurations of the 3-port network are shown.

From the previous problem and assuming that the indefinite scattering matrix is given by

$$\mathbf{S} = \begin{bmatrix} S_{11} & S_{12} & S_{13} \\ S_{21} & S_{22} & S_{23} \\ S_{31} & S_{32} & S_{33} \end{bmatrix}$$

we can write

$$S_{11}{}^{C} = S_{11}(0,0,-1) = \frac{S_{11} + \Delta_{22}}{1 + S_{33}}, \quad S_{12}{}^{C} = S_{12}(0,0,-1) = \frac{S_{12} - \Delta_{21}}{1 + S_{33}},$$

$$S_{21}{}^{C} = S_{21}(0,0,-1) = \frac{S_{21} - \Delta_{12}}{1 + S_{33}}, \quad S_{22}{}^{C} = S_{22}(0,0,-1) = \frac{S_{22} + \Delta_{11}}{1 + S_{33}}$$

where we have set $\Gamma_3 = -1$, as this is the port that is short circuited. In addition we have the following five relations between the indefinite scattering parameters (Equations (7.5.1) and (7.5.2)):

$$S_{13} = 1 - S_{11} - S_{12}, \qquad S_{23} = 1 - S_{21} - S_{22},$$

$$S_{31} = 1 - S_{11} - S_{21}, \qquad S_{32} = 1 - S_{12} - S_{22},$$

$$S_{33} = S_{11} + S_{12} + S_{21} + S_{22} - 1.$$

Substituting for Δ_{11}, Δ_{12}, Δ_{21} and Δ_{22} as

$$\Delta_{11} = S_{22}S_{33} - S_{23}S_{32} = \Delta_S - 1 + S_{22} + S_{12} + S_{21},$$

$$\Delta_{12} = S_{23}S_{31} - S_{21}S_{33} = \Delta_S + 1 - S_{21} - S_{22} - S_{11},$$

$$\Delta_{21} = S_{13}S_{32} - S_{12}S_{33} = \Delta_S + 1 - S_{12} - S_{11} - S_{22},$$

$$\Delta_{22} = S_{11}S_{33} - S_{13}S_{31} = \Delta_S - 1 + S_{11} + S_{12} + S_{21}$$

we have

$$S_{11}{}^C = \frac{\Delta_S - 1 + S_{12} + S_{21} + 2S_{11}}{S_{11} + S_{12} + S_{21} + S_{22}},$$

$$S_{12}{}^C = \frac{-\Delta_S - 1 + S_{11} + 2S_{12} + S_{22}}{S_{11} + S_{12} + S_{21} + S_{22}},$$

$$S_{21}{}^C = \frac{-\Delta_S - 1 + S_{11} + 2S_{21} + S_{22}}{S_{11} + S_{12} + S_{21} + S_{22}},$$

$$S_{22}{}^C = \frac{\Delta_S - 1 + S_{12} + S_{21} + 2S_{22}}{S_{11} + S_{12} + S_{21} + S_{22}}.$$

The above equations can be solved for S_{11}, S_{12}, S_{21} and S_{22} in terms of $S_{11}{}^C$, $S_{12}{}^C$, $S_{21}{}^C$ and $S_{22}{}^C$, giving

$$S_{11} = \frac{\Delta_S{}^C - 1 - 2S_{11}{}^C + S_{12}{}^C + S_{21}{}^C}{S_{11}{}^C + S_{12}{}^C + S_{21}{}^C + S_{22}{}^C - 4}, \qquad S_{21} = \frac{-\Delta_S{}^C - 1 - 2S_{21}{}^C + S_{22}{}^C + S_{11}{}^C}{S_{11}{}^C + S_{12}{}^C + S_{21}{}^C + S_{22}{}^C - 4},$$

$$S_{12} = \frac{-\Delta_S{}^C - 1 - 2S_{12}{}^C + S_{11}{}^C + S_{22}{}^C}{S_{11}{}^C + S_{12}{}^C + S_{21}{}^C + S_{22}{}^C - 4}, \qquad S_{22} = \frac{\Delta_S{}^C - 1 - 2S_{22}{}^C + S_{12}{}^C + S_{21}{}^C}{S_{11}{}^C + S_{12}{}^C + S_{21}{}^C + S_{22}{}^C - 4}$$

where

$$\Delta_S{}^C = S_{11}{}^C S_{22}{}^C - S_{12}{}^C S_{21}{}^C.$$

From the above expression and the previous relations between the indefinite scattering parameters, we can find the remaining indefinite scattering parameters as

$$S_{13} = \frac{-2 + 2S_{11}{}^C + 2S_{12}{}^C}{S_{11}{}^C + S_{12}{}^C + S_{21}{}^C + S_{22}{}^C - 4},$$

$$S_{23} = \frac{-2 + 2S_{22}{}^C + 2S_{21}{}^C}{S_{11}{}^C + S_{12}{}^C + S_{21}{}^C + S_{22}{}^C - 4},$$

$$S_{31} = \frac{-2 + 2S_{11}{}^C + 2S_{21}{}^C}{S_{11}{}^C + S_{12}{}^C + S_{21}{}^C + S_{22}{}^C - 4},$$

$$S_{32} = \frac{-2 + 2S_{22}{}^C + 2S_{12}{}^C}{S_{11}{}^C + S_{12}{}^C + S_{21}{}^C + S_{22}{}^C - 4},$$

$$S_{33} = \frac{-S_{11}{}^C - S_{12}{}^C - S_{21}{}^C - S_{22}{}^C}{S_{11}{}^C + S_{12}{}^C + S_{21}{}^C + S_{22}{}^C - 4}.$$

Once the indefinite scattering parameters are found, the expressions for Common B and Common A scattering parameters can easily be derived. These are given by the following expressions (see Example(7.3.2)):

$$S_{11}{}^B = S_{11}(0,-1,0) = \frac{S_{11} + \Delta_{33}}{1 + S_{22}},$$

$$S_{13}{}^B = S_{13}(0,-1,0) = \frac{S_{13} + \Delta_{31}}{1 + S_{22}},$$

$$S_{31}{}^B = S_{31}(0,-1,0) = \frac{S_{31} + \Delta_{13}}{1 + S_{22}},$$

$$S_{33}{}^B = S_{33}(0,-1,0) = \frac{S_{33} + \Delta_{11}}{1 + S_{22}},$$

$$S_{22}{}^A = S_{22}(-1,0,0) = \frac{S_{22} + \Delta_{33}}{1 + S_{11}},$$

$$S_{23}{}^A = S_{23}(-1,0,0) = \frac{S_{23} + \Delta_{32}}{1 + S_{11}},$$

$$S_{32}{}^A = S_{32}(-1,0,0) = \frac{S_{32} + \Delta_{23}}{1 + S_{11}},$$

$$S_{33}{}^A = S_{33}(-1,0,0) = \frac{S_{33} + \Delta_{22}}{1 + S_{11}}.$$

For conversion of common B to common A and C or the conversion of A to B and C, we can use the above expressions by renaming and renumbering the different ports.

Example (7.5.3)

The scattering parameters of a transistor in Common Emitter configuration are given as

$$S_{11}{}^{CE} = 0.36\angle -80°, \qquad S_{12}{}^{CE} = 0.07\angle 70°,$$

$$S_{21}{}^{CE} = 3.80\angle 96°, \qquad S_{22}{}^{CE} = 0.84\angle -9°.$$

Find the scattering parameters of the network in Common Collector and Common Base configurations.

Solution

Using the results of the previous example with designation of A to Emitter , B to Collector and C to Base, we first form the indefinite scattering parameters. With

$$\Delta_S{}^{CE} = S_{11}{}^{CE} S_{22}{}^{CE} - S_{12}{}^{CE} S_{22}{}^{CE} = 0.263 - j0.367,$$

we find the indefinite scattering parameters as

$$\mathbf{S} = \begin{bmatrix} 0.785 - j0.446 & 0.026 + j0.097 & 0.189 + j0.348 \\ -1.165 + j1.081 & 0.949 - j0.159 & 1.216 - j0.922 \\ 1.380 - j0.636 & 0.025 + j0.062 & -0.405 + j0.574 \end{bmatrix}.$$

With $\Delta_{11} = -0.380 + j0.397$, $\Delta_{13} = -1.304 + j0.778$, $\Delta_{31} = -0.113 + j0.206$ and $\Delta_{33} = 0.810 + j0.463$, we find

$$S_{11}{}^{CC} = 0.85 - j0.40, \qquad S_{13}{}^{CC} = 0.13 - j0.30,$$

$$S_{31}{}^{CC} = 1.43 - j0.61, \qquad S_{33}{}^{CC} = 0.45 - j0.54.$$

Similarly with $\Delta_{22} = -0.544 - j0.271$, $\Delta_{23} = -0.051 - j0.00$, $\Delta_{32} = -1.140 - j1.064$ and $\Delta_{33} = 0.810 - j0.463$, we find

$$S_{22}{}^{CB} = 1.01 - j0.10, \qquad S_{23}{}^{CB} = 1.50 - j0.74,$$

$$S_{32}{}^{CB} = -0.01 - j0.01, \qquad S_{33}{}^{CB} = -0.61 - j0.32.$$

7.6 Impedance Transformation and Matching

An N-port network, similar to a 2-port network, acts as an impedance transformer, changing the termination impedance of one port to the input impedance at a different port. In this case, however, the impedance transformation is affected also by the termination impedances of all other ports.

The input reflection coefficient of the jth port of an N-port network is defined as

$$\Gamma_{IN,j} = S_{jj}(\Gamma_1,.. \gamma_j = 0,.. \Gamma_N) \tag{7.6.1}$$

with $Z_{IN,j}$ related to $\Gamma_{IN,j}$, by the expression

$$Z_{IN,j} = \frac{1 + \Gamma_{IN,j}}{1 - \Gamma_{IN,j}} R_{0j}.$$

The input reflection coefficients or impedances are hence dependent on all network terminations, apart from the impedance termination of the input port.

The simultaneous consideration of the effects of all network terminations on a particular input reflection coefficient is, in general, complicated. We may, however, revert to our discussion of the last section and consider the N-port network, having an input port and an output port and $N - 2$ control ports. The impedance transformation properties of the equivalent 2-port network can then be considered for any particular set of the impedances of the control ports.

Without loss of generality, we can designate the input port as port 1 and the output port as port 2. If the reflection coefficient of port 2 is given by Γ_L, then we have

$$\Gamma_{IN,1} = S_{11}(0,\Gamma_L,.. \Gamma_j .. \Gamma_N) = S_{11}' + \frac{\Gamma_L S_{12}' S_{21}'}{1 - \Gamma_L S_{22}'}$$

where

$$S_{11}' = S_{11}(0, 0, .. \Gamma_j .. \Gamma_N), \tag{7.6.3a}$$

$$S_{12}' = S_{12}(0, 0, .. \Gamma_j .. \Gamma_N), \tag{7.6.3b}$$

$$S_{21}' = S_{21}(0, 0, .. \Gamma_j .. \Gamma_N), \tag{7.6.3c}$$

$$S_{22}' = S_{22}(0, 0, .. \Gamma_j .. \Gamma_N). \tag{7.6.3d}$$

The variation of $\Gamma_{IN,1}$ with Γ_L can be studied as in Chapter 3.
The scattering parameters S_{11}', S_{12}', S_{21}' and S_{22}' can be evaluated by setting $\Gamma_1 = 0$ and $\Gamma_2 = 0$ in (7.3.8).

Conjugate matching

The jth port of an N-port network is conjugate matched to the source if

$$\Gamma_j = S_{jj}^*(\Gamma_1, .. \Gamma_j = 0, .. \Gamma_N). \tag{7.6.4}$$

With (7.6.4) satisfied,

$$S_{jj}(\Gamma_1, \Gamma_2, .. \gamma_j = \Gamma_j, .. \Gamma_N) = 0. \tag{7.6.5}$$

Equivalent 2-port simultaneous conjugate matching

We may be interested to consider the simultaneous matching of two specific ports, with all other control port impedances fixed. This again reduces the problem to that of a 2-port network. To find Γ_1 and Γ_2 we can use the appropriate expressions of Chapter 3, with S_{11}, S_{12}, S_{21} and S_{22} replaced by S_{11}', S_{12}', S_{21}' and S_{22}' as given by (7.6.3).

N-port total simultaneous conjugate matching

For total simultaneous matching of all ports, we require that the following expressions to be simultaneously satisfied:

$$S_{11}(\Gamma_{1M} .. \Gamma_{jM} .. \Gamma_{NM}) = 0,$$
$$S_{22}(\Gamma_{1M} .. \Gamma_{jM} .. \Gamma_{NM}) = 0,$$
$$.............................$$
$$S_{jj}(\Gamma_{1M} .. \Gamma_{jM} .. \Gamma_{NM}) = 0,$$
$$.............................$$
$$S_{NN}(\Gamma_{1M} .. \Gamma_{jM} .. \Gamma_{NM}) = 0$$

or the diagonal terms of the generalized scattering matrix should be all zero.
In terms of the measured scattering parameters, this leads to N simultaneous equations of power N with N sets of solutions. Some set of solutions, however, may not be realizable with passive terminations. If a single port is generating, the simultaneous conjugate matching of the network leads, for some solutions, to maximum power derived from the source and minimum power dissipated in the network.

Example (7.6.1)

The scattering matrix of a 3-port network in a system of references $\hat{Z}_1 = \hat{Z}_2 = \hat{Z}_3 = 50$, is given by

$$
S = \begin{bmatrix}
0.367 - j0.479 & 0.293 + j0.097 & 0.070 + j0.625 \\
0.415 + j0.219 & -0.003 + j0.573 & 0.515 - j0.597 \\
0.097 + j0.243 & 0.702 + j0.062 & 0.773 + j0.419
\end{bmatrix}.
$$

If port 2 and 3 of the network are terminated by impedances $Z_2 = 100 + j50$ and $Z_3 = 50 - j100$, find the required source impedance for the input port (port 1), to conjugate match the network to this port.

Solution

The required reflection coefficient for the source at the input port is given by

$$
\Gamma_S = S_{11}{}^*(0, \Gamma_2, \Gamma_3).
$$

With $Z_1 = 50$, $Z_2 = 100 + j50$ and $Z_3 = 50 - j100$, we have $\Gamma_1 = 0$, $\Gamma_2 = 0.4 + j0.2$ and $\Gamma_3 = 0.5 - j0.5$. Using the conversion expression (7.3.8) we have

$$
S(0, \Gamma_2, \Gamma_3) = \begin{bmatrix}
0.614 - j0.071 & 0.819 - j0.112 & 0.255 + j1.059 \\
0.441 - j0.257 & -0.505 + j0.455 & 0.586 - j0.841 \\
0.220 + j0.609 & 0.916 - j0.014 & 0.972 + j0.655
\end{bmatrix}
$$

and hence $S_{11}(0, \Gamma_2, \Gamma_3) = 0.61 - j0.07$ or $\Gamma_S = 0.61 + j0.07$.
It can easily be verified that $S_{11}(\Gamma_S, \Gamma_2, \Gamma_3) = 0$

Example (7.6.2)

For the 3-port network of the previous example, show that the simultaneous conjugate matching can be achieved by terminating the network at port 1, port 2 and port 3 by impedances $Z_1 = 50 + j100$, $Z_2 = 25 - j50$ and $Z_3 = 100 - j25$, respectively.

Solution

The reflection coefficients of the terminations for ports 1, 2 and 3 are given, respectively, as $\Gamma_1 = 0.5 + j0.5$, $\Gamma_2 = 0.077 - j0.615$ and $\Gamma_3 = 0.351 - j0.108$. The scattering matrix for the network in a system of reference impedances equal to these impedances can be written as

$$\mathbf{S}(\Gamma_1, \Gamma_1, \Gamma_1) = \begin{bmatrix} 0 & 0.6 + j0.3 & 0.4 + j0.5 \\ 0.8 + j0.1 & 0 & 0.7 - j0.2 \\ 0.4 - j0.3 & 0.5 + j0.9 & 0 \end{bmatrix}$$

with all diagonal elements zero. We have, therefore, a 3-port simultaneous matching of the network.

7.7 *N*-Port Network Stability

The *j*th port of an *N*-port network oscillates if

$$Z_{Lj} + Z_{IN,j} = 0. \tag{7.7.1}$$

Substituting

$$Z_{Lj} = \left[\frac{1 + \Gamma_j}{1 - \Gamma_j}\right] R_{0j} \text{ and } Z_{IN,j} = \left[\frac{1 + \Gamma_{IN,j}}{1 - \Gamma_{IN,j}}\right] R_{0j}$$

into (7.7.1), we have

$$\Gamma_{Lj} \Gamma_{IN,j} = 1 \tag{7.7.2}$$

as an alternative condition for the oscillation of the *j*th port.
    Provided that

$$S_{jk}' = S_{jk}(\Gamma_1, .. \Gamma_j = 0, .. \Gamma_k = 0, .. \Gamma_N) \neq 0$$

and

$$S_{kj}' = S_{kj} (\Gamma_1, .. \Gamma_j = 0, .. \Gamma_k = 0, .. \Gamma_N) \neq 0,$$

then if (7.7.2) is satisfied, we have also

$$\Gamma_{Lk} \Gamma_{IN,k} = 1 \qquad (7.7.3)$$

or the kth port also oscillates.

An N-port network is stable at its jth port if

$$Re(Z_{Lj} + Z_{IN,j}) > 0 \qquad (7.7.4)$$

where Z_{Lj} is the load impedance of the jth port and $Z_{IN,j}$, the input impedance of the N-port network as seen from the same port.

When

$$Re(Z_{Lj} + Z_{IN,j}) < 0, \qquad (7.7.5)$$

the circuit is unstable and will oscillate as explained in Chapter 4.

Network stability for all load impedances of a particular port

Expression (7.7.4), gives the condition for the stability of the jth port of an N-port network for a particular value of Z_{Lj}. We can further consider the stability of the jth port for *all* possible passive values of Z_{Lj}. This is ensured if

$$Re(Z_{IN,j}) > 0. \qquad (7.7.6)$$

Substituting for

$$Re(Z_{IN,j}) = \frac{1}{2}(Z_{IN,j} + Z_{IN,j}{}^*)$$

and for $Z_{IN,j}$ in terms of $\Gamma_{IN,j}$, we can write (7.7.6) as

$$\frac{1 - |\Gamma_{IN,j}|^2}{|1 - \Gamma_{IN,j}|^2} > 0$$

or

$$|\Gamma_{IN,j}| = |S_{jj}(\Gamma_1, .. \Gamma_j = 0, .. \Gamma_N)| < 1. \qquad (7.7.7)$$

Hence an N-port network terminated at its jth port by any load impedance will be stable if the magnitude of the input reflection coefficient at that port is less than unity.

Equivalent 2-port *unconditional stability*

For convenience let us call the relevant ports as port 1 and port 2. If all other ports are terminated by fixed impedances, the equivalent 2-port network is unconditionally stable, if it is stable for all values of Z_{L1}, and Z_{L2}.

Denoting

$$S_{11}' = S_{11}(0,0, .. \Gamma_j .. \Gamma_N),$$

$$S_{12}' = S_{12}(0,0, .. \Gamma_j .. \Gamma_N),$$

$$S_{21}' = S_{21}(0,0, .. \Gamma_j .. \Gamma_N),$$

$$S_{22}' = S_{22}(0,0, .. \Gamma_j .. \Gamma_N)$$

the unconditional stability of the equivalent 2-port network is identical to that of any 2-port network, provided that we replace S_{11}, S_{12}, S_{21} and S_{22} by $S_{11}', S_{12}', S_{21}'$ and S_{22}' respectively. The unconditional stability is hence ensured if

$$K' = \frac{1 - |S_{11}'|^2 - |S_{22}'|^2 + |\Delta_S'|^2}{2|S_{12}' S_{21}'|} > 1 \qquad (7.7.9)$$

and

$$|\Delta_S'| = |S_{11}' S_{22}' - S_{12}' S_{21}'| < 1. \qquad (7.7.10)$$

The effect of each termination Z_{Lj} ($j \neq 1$, $j \neq 2$) on the unconditional stability of the network can be studied, by mapping of K' and $|\Delta_S'|$ values in a Γ_{Lj} plane.

Example (7.7.1)

Derive the unconditional stability factors $|\Delta_S'|$ and K' for ports 1 and 2 of a 3-port network, expressed in terms of the measurable scattering parameters and the reflection coefficient Γ_3 of port 3.

Solution

Δ_S' and K' can be found from (7.7.9) and (7.7.10) and by the following substitutions

$$S_{11}' = \frac{S_{11} - \Gamma_3 \Delta_{22}}{1 - \Gamma_3 S_{33}}, \quad S_{12}' = \frac{S_{12} + \Gamma_3 \Delta_{21}}{1 - \Gamma_3 S_{33}},$$

$$S_{21}' = \frac{S_{21} + \Gamma_3 \Delta_{12}}{1 - \Gamma_3 S_{33}}, \quad S_{22}' = \frac{S_{22} - \Gamma_3 \Delta_{11}}{1 - \Gamma_3 S_{33}}$$

giving

$$\left| \Delta_S' \right| = \left| \frac{\Delta_{33} - \Gamma_3 \Delta}{1 - \Gamma_3 S_{33}} \right|$$

and

$$K' = \frac{E \Gamma_3 \Gamma_3^* - F^* \Gamma_3 - F \Gamma_3^* + G}{2 \left| H \Gamma_3^2 + L \Gamma_3 + M \right|}$$

where

$$E = \left| S_{33} \right|^2 - \left| \Delta_{22} \right|^2 - \left| \Delta_{11} \right|^2 + \left| \Delta \right|^2, \qquad H = \Delta_{12} \Delta_{21},$$

$$F = S_{33}^* - \Delta_{22}^* S_{11} - \Delta_{11}^* S_{22} + \Delta^* \Delta_{33}, \qquad L = S_{21} \Delta_{21} + S_{12} \Delta_{12},$$

$$G = 1 - \left| S_{11} \right|^2 - \left| S_{22} \right|^2 + \left| \Delta_{33} \right|^2, \qquad M = S_{12} S_{21}.$$

Example (7.7.2)

Examine the equivalent 2-port unconditional stability of a 3-port network with the S-matrix

$$S = \begin{bmatrix} 0.648 + j0.297 & -0.073 - j0.449 & 0.034 + j0.150 \\ 0.011 - j0.422 & 0.541 + j0.359 & -0.134 + j0.241 \\ -0.119 + j0.186 & -0.228 + j0.456 & 0.645 - j0.416 \end{bmatrix}$$

as given in Example (7.3.3). Take port 1 and port 2 as the input and output ports and port 3 as the control port terminated by impedances (a) $Z_3 = 25 + j75$ and (b) $Z_3 = 75 - j25$.

Solution

(a) The required parameters can be found as from the previous example as

$\Gamma_3 = 0.333 + j0.667,$

$\Delta_{11} = 0.578 + j0.123, \quad \Delta_{22} = 0.573 - j0.066, \quad \Delta_{33} = 0.434 + j0.367,$

$\Delta = 0.432 + j0.222,$

$E = 0.143, \quad H = -0.032 + j0.069, \quad F = 0.206 + j0.124, \quad L = 0.193 - j0.143,$

$G = 0.394, \quad M = -0.190 + j0.026, \left| \Delta_S' \right| = 0.748, \quad K' = 0.987.$

The above results can also be found by finding the equivalent 2-port scattering parameters as

$$S_{11}' = 0.665 + j0.258, \qquad S_{12}' = -0.044 - j0.544,$$

$$S_{21}' = 0.081 - j0.454, \qquad S_{22}' = 0.695 + j0.266$$

and using the usual expressions for $\left| \Delta_S \right|$ and K for a 2-port network. For a load of $Z_3 = 25 + j75$, therefore, the 3-port network is not unconditionally stable($K' < 1$).

(b) For this case $\Gamma_3 = 0.231 - j0.154$. $\Delta_{11}, \Delta_{22}, \Delta_{33}$, Δ, E, F, G, H, L and M, however, have the same value as in (a). Substituting into the expressions for $\left| \Delta_S' \right|$ and K' as given in previous Example, we find $\left| \Delta_S' \right| = 0.52$ and $K' = 1.04$. Hence with $\left| \Delta_S' \right| < 1$ and $K' > 1$, the 3-port network is unconditionally stable for $Z_3 = 75 - j25$.

7.8 *N*-Port Network Power Considerations

Consider an *N*-port network as shown in Fig. (7.8.1) with the scattering parameters given in a system of reference impedances

$$\hat{Z}_j = Z_j \left(\gamma_j = \Gamma_j \right)$$

where Z_j is the load or source impedance of the *j*th port.

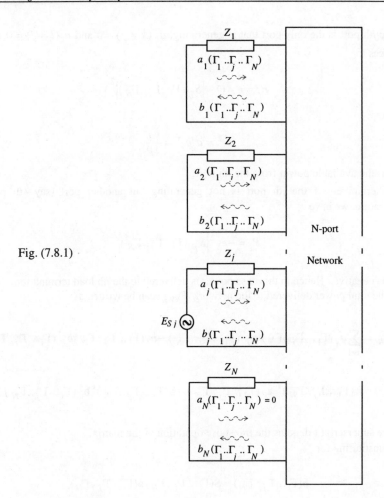

Fig. (7.8.1)

Power delivered to the *j*th port is given by the expression

$$P_j = a_j(\Gamma_1,..\Gamma_j..\Gamma_N)\, a_j^*(\Gamma_1,..\Gamma_j..\Gamma_N) - b_j(\Gamma_1,..\Gamma_j..\Gamma_N)\, b_j^*(\Gamma_1,..\Gamma_j..\Gamma_N)$$

$$= a_j(\Gamma_1,..\Gamma_j..\Gamma_N)\, a_j^*(\Gamma_1,..\Gamma_j..\Gamma_N) -$$

$$\sum_{k=1}^{N}\sum_{l=1}^{N} S_{jk}(\Gamma_1,..\Gamma_j..\Gamma_N)\, S_{jl}^*(\Gamma_1,..\Gamma_j..\Gamma_N)\, a_k(\Gamma_1,..\Gamma_j..\Gamma_N)\, a_l^*(\Gamma_1,..\Gamma_j..\Gamma_N).$$

If the jth port is the only port that is generating, $a_k\,(k \neq j) = 0$ and $a_l\,(l \neq j) = 0$ and P_j reduces to

$$P_j = e_j^2 (1 - \left| S_{jj}(\Gamma_1..\Gamma_j..\Gamma_N) \right|^2) \tag{7.8.2}$$

where

$$e_j^2 = a_j a_j^* = \frac{E_j^2}{4R_j}$$

and is the available power from the jth port.

Alternatively if the jth port is not generating but another port (say kth port) is generating, we have

$$P_j = -e_k^2 \left| S_{jk}(\Gamma_1..\Gamma_j..\Gamma_N) \right|^2 \tag{7.8.3}$$

and is negative. Hence in this case power is delivered to the jth load termination.

The total power delivered to the network P_{NWK} can be written as

$$P_{NWK} = \sum_{j=1}^{N} a_j(\Gamma_1,..\Gamma_j..\Gamma_N)\,a_j^*(\Gamma_1,..\Gamma_j..\Gamma_N) - b_j(\Gamma_1,..\Gamma_j..\Gamma_N)b_j^*(\Gamma_1,..\Gamma_j..\Gamma_N)$$

$$= \mathbf{a}(\Gamma_1,..\Gamma_j..\Gamma_N)^t\,\mathbf{a}^*(\Gamma_1,..\Gamma_j..\Gamma_N) - \mathbf{b}(\Gamma_1,..\Gamma_j..\Gamma_N)^t\mathbf{b}^*(\Gamma_1,..\Gamma_j..\Gamma_N)$$

where superscript $\mathbf{t}$ denotes the transpose operation of the matrix.

Substituting for

$$\mathbf{b}(\Gamma_1,..\Gamma_j..\Gamma_N) = \mathbf{S}(\Gamma_1,..\Gamma_j..\Gamma_N)\,\mathbf{a}(\Gamma_1,..\Gamma_j..\Gamma_N),$$

we have

$$P_{NWK} = \mathbf{a}(\Gamma_1,..\Gamma_j..\Gamma_N)^t\,\mathbf{Q}(\Gamma_1,..\Gamma_j..\Gamma_N)\,\mathbf{a}^*(\Gamma_1,..\Gamma_j..\Gamma_N)$$

where

$$\mathbf{Q}(\Gamma_1,..\Gamma_j..\Gamma_N) = \mathbf{U} - \mathbf{S}(\Gamma_1,..\Gamma_j..\Gamma_N)^t\mathbf{S}^*(\Gamma_1,..\Gamma_j..\Gamma_N).$$

The above expression for the total power delivered to the network is independent of the assumed reference impedances and hence taking $\hat{Z}_j = R_{0j}$ $(j = 1,.., N)$ the net power remains unchanged. Hence the power delivered to the network can be simply written as

$$P_{NWK} = \mathbf{a}^t \, \mathbf{Q} \, \mathbf{a}^* \tag{7.8.4}$$

where

$$\mathbf{Q} = \mathbf{U} - \mathbf{S}^t \, \mathbf{S}^*. \tag{7.8.5}$$

The matrix $\mathbf{Q}$ has the special property of being equal to its transposed conjugate. This is simply verified by conjugating and then transposing the right-hand side of (7.8.5) which will remain unchanged. The matrix $\mathbf{Q}$ is hence Hermitian, with $Q_{jk} = Q^*_{kj}$ and all diagonal elements real.

Passive Networks

For a passive N-port network,

$$P_{NWK} \geq 0$$

for all passive terminations. Hence we can write

$$P_{NWK} = \mathbf{a}^t \, \mathbf{Q} \, \mathbf{a}^* \geq 0. \tag{7.8.6}$$

As $\mathbf{Q}$ is Hermitian, (7.8.6) is satisfied if all determinants of the submatrices of $\mathbf{Q}$ are zero or positive. Hence the necessary and sufficient condition for a network to be passive is that the determinant of the submatrices of the matrix

$$\mathbf{Q} = \mathbf{U} - \mathbf{S}^t \, \mathbf{S}^*$$

be non-negative.

Lossless networks

An N-port network is lossless if

$$P_{NWK} = 0$$

for all possible terminations. Hence for an N-port network to be a lossless, we have

$$\mathbf{a}^t \, \mathbf{Q} \, \mathbf{a}^* = 0 \tag{7.8.7}$$

for all $\mathbf{a}$.

As (7.8.7) has to be satisfied for all **a**, we have

$$Q = 0$$

or

$$S^t S^* = U. \tag{7.8.8}$$

Reciprocal networks

For a network to be reciprocal we require

$$Z = Z^t \tag{7.8.9}$$

where **Z** is the impedance matrix of the network and **t** denotes the transpose operation of the matrix.

Equation (7.3.1) can be used to express **Z** in terms of **S** in a system of resistive reference $\hat{Z}_j = R_{0j}$ ($j = 1,.., N$). Hence

$$Z \equiv \{U - S\}^{-1}\{U + S\}[R_0]$$

$$= Z^t \equiv [R_0]\left(\{U - S\}^{-1}\{U + S\}\right)^t$$

or

$$\{U - S\}^{-1}\{U + S\} = \{U + S^t\}\{U - S^t\}^{-1}.$$

Pre-multiplying both sides by $\{U - S\}$ and post-multiplying by $\{U - S^t\}$, we have

$$\{U - S\}\{U + S^t\} = \{U + S\}\{U - S^t\}$$

which reduces to

$$S = S^t \tag{7.8.10}$$

or $S_{kj} = S_{jk}$ as the condition for the reciprocity of an *N*-port network.

Example (7.8.1)

Show that the conditions given for an *N*-port network to be passive, lossless and reciprocal, reduces to those given in Section (5.5) for a 2-port network.

Solution

For a 2-port network,

$$
\mathbf{Q} = \mathbf{U} - \mathbf{S}^t\,\mathbf{S}^* =
\begin{bmatrix} 1 & 0 \\ 0 & 1 \end{bmatrix}
-
\begin{bmatrix} S_{11} & S_{21} \\ S_{12} & S_{22} \end{bmatrix}
\begin{bmatrix} S_{11}{}^* & S_{12}{}^* \\ S_{21}{}^* & S_{22}{}^* \end{bmatrix}
$$

$$
=
\begin{bmatrix}
1 - |S_{11}|^2 - |S_{21}|^2 & S_{11}S_{12}{}^* + S_{21}S_{22}{}^* \\
S_{12}S_{11}{}^* + S_{22}S_{21}{}^* & 1 - |S_{22}|^2 - |S_{12}|^2
\end{bmatrix}.
$$

As it can be seen $\mathbf{Q}$ is Hermitian with $Q_{12} = Q_{21}{}^*$ and all diagonal elements are real. For the network to be passive, the determinants of all submatrices of $\mathbf{Q}$ should be positive. Hence

$$
|S_{11}|^2 + |S_{21}|^2 < 1,
$$

$$
|S_{22}|^2 + |S_{12}|^2 < 1,
$$

$$
(|S_{11}|^2 + |S_{21}|^2 - 1)\ (|S_{22}|^2 + |S_{12}|^2 - 1)
$$

$$
- (S_{12}S_{11}{}^* + S_{22}S_{21}{}^*)\ (S_{11}S_{12}{}^* + S_{21}S_{22}{}^*)
$$

$$
= |\Delta_S|^2 - |S_{11}|^2 - |S_{12}|^2 - |S_{21}|^2 - |S_{22}|^2 + 1 > 0
$$

as found in Section (5.5).

For the 2-port network to be lossless, we require $\mathbf{Q} = \mathbf{0}$, or

$$
|S_{11}|^2 + |S_{21}|^2 = 1, \qquad S_{11}S_{12}{}^* + S_{21}S_{22}{}^* = 0,
$$

$$
|S_{22}|^2 + |S_{12}|^2 = 1, \qquad S_{12}S_{11}{}^* + S_{22}S_{21}{}^* = 0
$$

leading to the results of Section (5.5).

For the 2-port network to be reciprocal we require $\mathbf{S} = \mathbf{S}^t$ or $S_{12} = S_{21}$ as found before.

Example (7.8.2)

Find the conditions for a 3-port network to be (i) lossless (ii) reciprocal.

Solution

(i) For a 3-port network to be lossless, we have

$$Q = U - S^t \, S^* = 0$$

or

$$\begin{bmatrix} 1 & 0 & 0 \\ 0 & 1 & 0 \\ 0 & 0 & 1 \end{bmatrix} - \begin{bmatrix} S_{11} & S_{21} & S_{31} \\ S_{12} & S_{22} & S_{32} \\ S_{13} & S_{23} & S_{33} \end{bmatrix} \begin{bmatrix} S_{11}^* & S_{12}^* & S_{13}^* \\ S_{21}^* & S_{22}^* & S_{23}^* \\ S_{31}^* & S_{32}^* & S_{33}^* \end{bmatrix} = \begin{bmatrix} 0 & 0 & 0 \\ 0 & 0 & 0 \\ 0 & 0 & 0 \end{bmatrix}.$$

Hence

$$Q_{11} = 1 - |S_{11}|^2 + |S_{21}|^2 + |S_{31}|^2 = 0,$$

$$Q_{12} = S_{11}S_{12}^* + S_{21}S_{22}^* + S_{31}S_{32}^* = 0,$$

$$Q_{13} = S_{11}S_{13}^* + S_{21}S_{23}^* + S_{31}S_{33}^* = 0,$$

$$Q_{21} = S_{12}S_{11}^* + S_{22}S_{21}^* + S_{32}S_{31}^* = 0,$$

$$Q_{22} = 1 - |S_{12}|^2 + |S_{22}|^2 + |S_{32}|^2 = 0,$$

$$Q_{23} = S_{12}S_{13}^* + S_{22}S_{23}^* + S_{32}S_{33}^* = 0,$$

$$Q_{31} = S_{13}S_{11}^* + S_{23}S_{21}^* + S_{33}S_{31}^* = 0,$$

$$Q_{32} = S_{13}S_{12}^* + S_{23}S_{22}^* + S_{33}S_{32}^* = 0,$$

$$Q_{33} = 1 - |S_{13}|^2 + |S_{23}|^2 + |S_{33}|^2 = 0.$$

However, as $\mathbf{Q}$ is Hermitian, with $Q_{12} = Q_{13} = Q_{23} = 0$, the conditions $Q_{21} = Q_{31} = Q_{32} = 0$ are redundant.

(ii) For a 3-port network to be reciprocal, we have

$$\begin{bmatrix} S_{11} & S_{12} & S_{13} \\ S_{21} & S_{22} & S_{23} \\ S_{31} & S_{32} & S_{33} \end{bmatrix}^{t} = \begin{bmatrix} S_{11} & S_{12} & S_{13} \\ S_{21} & S_{22} & S_{23} \\ S_{31} & S_{32} & S_{33} \end{bmatrix}$$

or $S_{12} = S_{21}$, $S_{13} = S_{31}$ and $S_{23} = S_{32}$.

Appendix

Conversion of the Scattering Matrices in Systems of Different Reference Impedances

The generalized scattering matrix in terms of the impedance matrix was given by (7.2.5) as

$$\mathbf{S}(\gamma_1 .. \gamma_j .. \gamma_N) = \left[\frac{1}{\sqrt{\widehat{R}}}\right]\{\mathbf{Z} - \widehat{\mathbf{Z}}^*\}\{\mathbf{Z} + \widehat{\mathbf{Z}}\}^{-1}\left[\sqrt{\widehat{R}}\right] \tag{A.1}$$

where from (7.3.3)

$$\hat{\mathbf{Z}} = [R_0]\{\mathbf{U} + \gamma\}\{\mathbf{U} - \gamma\}^{-1} \tag{A.2}$$

and the impedance matrix in terms of the measured scattering matrix by (7.3.1) as

$$\mathbf{Z} = [R_0]\{\mathbf{U} - \mathbf{S}\}^{-1}\{\mathbf{U} + \mathbf{S}\}. \tag{A.3}$$

From (7.3.6) and (7.3.7), we have respectively

$$\left[\sqrt{\hat{R}}\right] = \left[\sqrt{R_0}\right]\{\mathbf{U} - \gamma\}^{-1}\Lambda^* \tag{A.4}$$

and

$$\left[\frac{1}{\sqrt{\hat{R}}}\right] = \left[\frac{1}{\sqrt{R_0}}\right]\Lambda^{-1}\{\mathbf{U} - \gamma^*\}. \tag{A.5}$$

Substituting for $\hat{\mathbf{Z}}, \mathbf{Z}$, $\left[\sqrt{\hat{R}}\right]$ and $\left[1/\sqrt{\hat{R}}\right]$ from (A.2) to (A.5) into (A.1), we have

$$\mathbf{S}(\gamma_1..\gamma_j..\gamma_N) = \left[\frac{1}{\sqrt{R_0}}\right] \ \Lambda^{-1}\{\mathbf{U} - \gamma^*\}$$

$$\times\left([R_0]\{\mathbf{U} - \mathbf{S}\}^{-1}\{\mathbf{U} + \mathbf{S}\} + [R_0]\{\mathbf{U} - \gamma^*\}^{-1}\{\mathbf{U} + \gamma^*\}\right)$$

$$\times\left([R_0]\{\mathbf{U} - \mathbf{S}\}^{-1}\{\mathbf{U} + \mathbf{S}\} - [R_0]\{\mathbf{U} - \gamma\}^{-1}\{\mathbf{U} + \gamma\}\right)^{-1}$$

$$\times\left[\sqrt{R_0}\right]\{\mathbf{U} - \gamma\}^{-1}\Lambda^*$$

$$= \Lambda^{-1}\left(\{\mathbf{U} - \gamma^*\}\{\mathbf{U} - \mathbf{S}\}^{-1}\{\mathbf{U} + \mathbf{S}\} - \{\mathbf{U} + \gamma^*\}\right)$$

$$\times\left(\{\mathbf{U} - \gamma\}\{\mathbf{U} - \mathbf{S}\}^{-1}\{\mathbf{U} + \mathbf{S}\} - \{\mathbf{U} + \gamma\}\right)^{-1}\Lambda^*.$$

With the substitution of

$$\{U + S\} = \{U - S\} + 2S,$$

the above expression can be written as

$$S(\gamma_1 \cdot \cdot \gamma_j \cdot \cdot \gamma_N) = \Lambda^{-1} \left(-\gamma^* + \{U - \gamma^*\}\{U - S\}^{-1}S \right)\left(U + \{U - \gamma\}\{U - S\}^{-1}S \right)^{-1} \Lambda^*.$$

However,

$$\left(-\gamma^* + \{U - \gamma^*\}\{U - S\}^{-1}S \right) = \left(-\gamma^*\{S^{-1} - U\} + \{U - \gamma^*\} \right)\{U - S\}^{-1}S$$

and

$$\left(U + \{U - \gamma\}\{U - S\}^{-1}S \right)^{-1} = S^{-1}\{U - S\}\left(U\{S^{-1} - U\} + \{U - \gamma\} \right)^{-1}.$$

Hence

$$S(\gamma_1 \cdot \cdot \gamma_j \cdot \cdot \gamma_N) = \Lambda^{-1} \left(-\gamma^*\{S^{-1} - U\} + \{U - \gamma^*\} \right)\left(U\{S^{-1} - U\} + \{U - \gamma\} \right)^{-1} \Lambda^*$$

$$= \Lambda^{-1}\{U - \gamma^*S^{-1}\}\{US^{-1} - \gamma\}^{-1}\Lambda^*$$

and finally

$$S(\gamma_1 \cdot \cdot \gamma_j \cdot \cdot \gamma_N) = \Lambda^{-1}\{S - \gamma^*\}\{U - \gamma S\}^{-1}\Lambda^*.$$

Index